INDIGENOUS RESOURCES FOR RURAL DEVELOPMENT

By the Same Author

- Education and Socio-economic Development: Rural Urban Divide in India and South Asia
- Globalization: Debts, Arms and Reforms in South Asia
- नौकरशाही का प्रबन्धः आर्थिक सुधारों एवं ग्रामीण विकास के लिए पुनर्गठन
- पुलिस एवं अदालतों में नागरिक सहभागिताः सस्ते-एवं-शीघ्र न्याय के लिए एक प्रशासनिक ढाँचा

About the Author

Before taking up a contractual assignment as Resident Consultant for one year with the N.P.I.U. for implementing the World Bank project under the Ministry of HRD, Government of India, **Dr. Sures Chandra Jain** had worked as Chief General Manager for nearly 20 years at M.P. Laghu Udyog Nigam (an undertaking of the Government of Madhya Pradesh) in Bhopal when he voluntarily retired in 2002. Dr. Jain had also worked for many multinational corporations like I.B.M., Massey-Ferguson, Atomic Energy of Canada Limited, and G.T.E. when he was in Canada and USA from 1970 to 1983. He was a visiting Fellow at UNESCO, Paris, France.

Dr. Jain obtained the degrees of Doctor of Education (Ph.D.) from Columbia University, New York, U.S.A. and B.I.S. from the University of Waterloo in Canada, Besides several research papers, Dr. Jain has written five books, two in Hindi and three in English, all of them published by Concept Publishing Company, New Delhi. He is either an author or co-author of many government reports which are not meant for general publication.

INDIGENOUS RESOURCES FOR RURAL DEVELOPMENT

Agricultural Mechanisation and Rural Industrialisation

Dr. Sures Chandra Jain

CONCEPT PUBLISHING COMPANY, NEW DELHI-110059

Cataloging in Publication Data—DK
Courtesy: D.K. Agencies (P) Ltd. <docinfo@dkagencies.com>

Jain, Sures Chandra, 1946-
Indigenous resources for rural development : agricultural mechanisation and rural industrialisation/ Sures Chandra Jain.
p. cm.
Includes bibliographical references.
Includes index.
ISBN 8180691527

1. Rural development—Effect of appropriate technology on—India. 2. Farm mechanization—India. 3. Rural industries—India. 4. Appropriate technology—India. I. Title.

307.141 209 54 21

ISBN : 81-8069-152-7

First Published 2005

Published and Printed by
Ashok Kumar Mittal
Concept Publishing Company
A/15-16, Commercial Block, Mohan Garden
New Delhi-110059 (INDIA)

Phone : 25351460, 25351794
Fax : 091-11-25357103
Email : publishing@conceptpub.com

Preface

Whether a person in India and in developing countries below the poverty line or above it, varies substantially from source to source, from institution to institution, and from country to country. To my mind, a person is below the level of poverty line when he or she is not in a financial position to afford the following five basic requirements of dignified human life:

1. Sufficient food,
2. Minimum hygienic housing,
3. Adequate clothing,
4. Access to primary education, and
5. Basic health care.

Within the limits of these five constraints, it would be safe to state that over 60 per cent of the population in urban areas, and over 80 per cent of the population in rural areas in India live below the poverty line. Approximately 80 per cent of the total population of India, 843 million in 1991, live in rural areas. The situation of dealing with poverty is very precarious because of the following reasons:

1. A person is poor because he has no skill to secure a gainful employment or create his own facility for self-employment.
2. He has no skill because, being poor, he cannot afford even primary education; he ends up working as a child labour.

So there is a real paradox here. This book is an attempt towards indigenous resource-based economic development, with reference to an analysis of economic inequality between urban centres and rural areas; evolution of a self-reliant, agro-industrial complex based on agricultural mechanisation, rural industrialisation, and utilisation of natural resources.

This book is an analysis of poverty *versus* polity, as it attempts to define the constraints and imperatives in the task of socio-economic transformation based on evolutionary, reformist and revolutionary strategies.

The Book is primarily designed as a general reference material for the layman who wishes to become familiar with the subject of agricultural mechanisation and rural industrialisation, as it relates with the overall socio-economic development of the country.

This book was originally written in 1982 when I was in Canada, updated and thoroughly revised ten years later in 1992, and now updated and revised again the same ten years later in 2002. As a result, some part of this book may in the first instance appear to be a bit disjointed but, in fact, analyses and arguments made 20 years ago and 10 year ago are even much more relevant today. This is precisely because poverty and rural development continue to be perennial problems and pressing issues facing India and other developing country. These problems of poverty and rural development have only gone from bad to worse in these last 20 years when the book was originally written.

I wish to express my appreciation to my wife, Dr. Krishna and our two little children, Nidhi and Amogh, who so gracefully provided me with what otherwise would have been their time.

Bhopal, **Sures Chandra Jain**
September 12, 2002

Contents

Contents

List of Tables

Introduction

Can an improvement in governance brought about by constitutional reform assist in the task of poverty alleviation, a task on which successive governments have failed to deliver for the last 53 years?... The lack of good governance according to me is an important reason behind persistent poverty. ...

—Dhiraj Nayyar[1], 2000

A poor man does not use money as a weapon. The major means of production are: the land, forests; mineral resources; water; oil and industry...upon which a large section of population depends for their living, or...which produce essential raw materials.

—Julius K. Nyerere[2], 1967

...Poor people are active agents in their lives, but are often powerless to influence the social and economic factors that determine their well-being. ...The world has deep poverty amid plenty. Of the world's 6 billion people (in the year 2000) 2.8 billion—almost half—live on less than $2 a day, and 1.2 billion—a fifth—live on less than $1 a day, with 44 per cent living in South Asia.

The World Bank[3], 2000/2001

While living under the general affluence of the West from 1970 to 1983, I often used to wonder as to what could be the causes of grinding poverty in India and the Third World. I eventually realised that one of the fundamental causes of the continuing poverty in India is that our polity, largely dictated by vested interests, has so far been with an urban bias, during almost the half a century since Independence in 1947. The polity is, in a nutshell, also broadly based on western-style modernisation, catering to the needs and aspirations of the upper strata of the society, and ignoring the genuine need for equality and social justice for nearly half of the population of the country, even after half-a-century-of-Independents which barely subsists either just at the poverty line or below it. The World Bank in its World Development Report 2000/2001 reported that:[4]

> "...Poor people are active agents in their lives, but are often powerless to influence the social and economic factors that determine their well-being. ... The world has deep poverty amid plenty. Of the world's 6 billion people (in the year 2000) 2.8 billion—almost half—live on less than $2 a day, and 1.2 billion—a fifth—live on less than $1 a day, with 44 per cent living in South Asia."

Approximately 80 per cent of 843 million people of India in 1991 live in more than half a million villages, most of them earning their livelihood at a bare subsistence level.

It is quite obvious that the task of nation-building in the post-Independence era since 1947 has, by and large, been wholly biased towards urban centres, with substantially based on foreign aid, debt and technology.[5] The disparities concerning income, education, agricultural mechanisation, and rural industrialisation, the revival of self-dignity to off-set the British colonial dehumanisation, degradation and enforced servility and flunkeyism, all amount to the near total neglect, deliberate in most cases, of rural areas. The main beneficiaries of the fruits of Independence have been approximately 10 per cent of the people, most of them in urban centres. The statistical analyses of the data in various chapters clearly prove these facts.

Lack of primary education[6] and poverty logically feed each other. In an attempt to remove poverty, the attack therefore has to be on both the fronts simultaneously. In this book, I am concerned with the programmes of action for the removal of poverty only for rural development, as it relates with agricultural mechanisation and rural industrialisation.

There are six chapters in this book:

Chapter 1: The Problem: People below the Poverty Line

This chapter is an analysis of the problem of poverty in India and the Third World. There is no point in saying that there has been less and less percentage of people below the poverty line in nearly half a century since Independence in 1947. To my mind, the fact is that a person is below the level of poverty line when he or she is not in a financial position to afford the following five basic requirements of dignified human life:

1. Sufficient food,
2. Minimum hygienic housing,
3. Adequate clothing,
4. Access to primary education, and
5. Basic health care.

The figures for the population living below the poverty line in India vary substantially from source to source, from institution to institution, and from economist to economist. But this concept of defining the poverty line is, "by any consideration, highly arbitrary", Vasant Sathe[7] said in 1988 when he was a Cabinet Minister. Sathe felt that over 80 per cent in rural areas and over 60 per cent of the population in urban centres live below the poverty line. Various official and other estimates of people below the poverty line are given in chapter 1 in the text.

The real beneficiaries of production and productivity in the post-Independence era have been approximately 10 per cent of the people; most of them in urban centres, while over 80 per cent of the population subsist in rural areas. The data are provided in tabular format, and analyzed for India and the Third World.

Chapter 2: The Formation of an Agro-Industrial Complex

In this chapter, I have attempted to devise a model of the sequences of change, which would lead to a relatively smooth

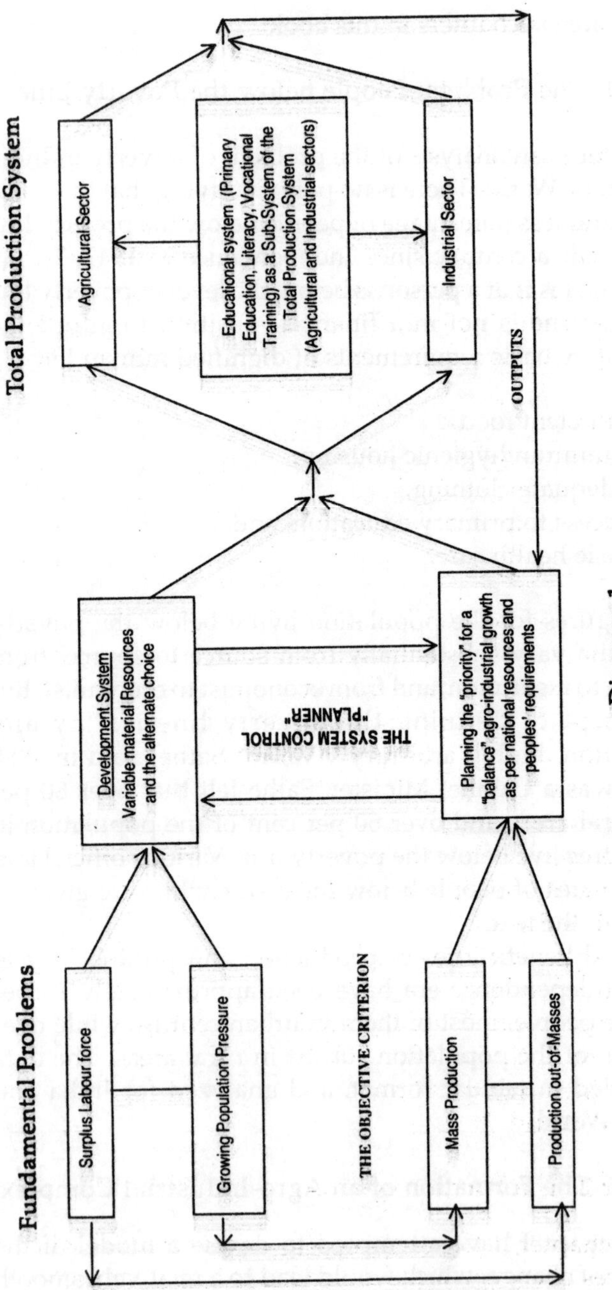

Diagram 1

Decision-making Variables for Agricultural Mechanisation and Rural Industrialisation

Flow Chart 1

Linking Agricultural Mechanisation and Rural Industrialisation for a Generalized Developing Economy

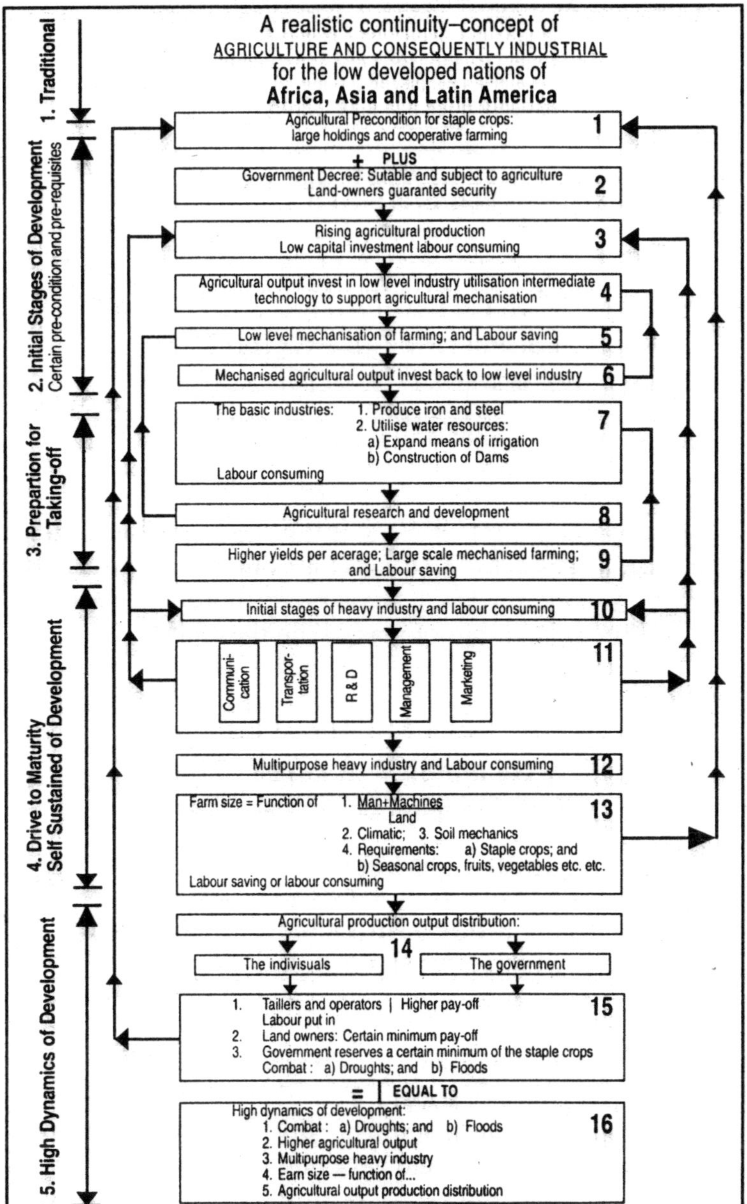

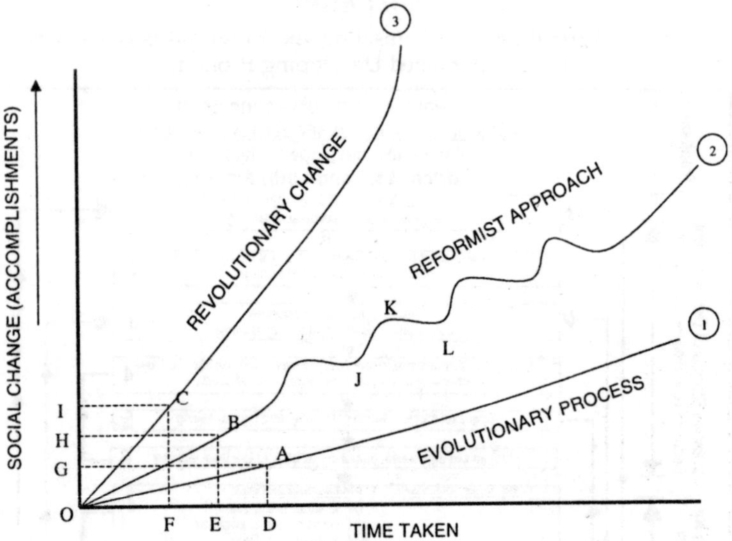

modernisation of agriculture and the subsequent emergence of industrialisation for rural development. In this model, I am particularly concerned with 'mass-production' on the one hand and the 'production-out-of-masses' on the other.

This model is based on certain policy assumptions related with:

1. Rural development: land owners and tillers.
2. Labour-intensive small-scale industries.

According to Mahatma Gandhi (1936)[8]:

> Real socialism has been handed down to us by our ancestors who taught: "All land belongs to Gopal, where then is the boundary line? Man is the maker of the line and he can therefore unmake it". Gopal literally means shepherd; it also means God. In modern language it means the State, *i.e.*, the people. That the land today does not belong to people is too true. But the fault is not in the teaching. It is in us who have not lived up to it.

In my attempt towards the formation of an agro-industrial complex for agricultural-mechanisation and rural industrialisation, I have made the following provisions in the model:

1. Tillers and operators with the highest pay-off in proportion of output, compared with the amount of labour input.
2. Land owners with a guaranteed security for their land and a certain minimum pay-off.
3. Incentives to both land owners and landless peasantry for their land and for their labour input, respectively.
4. Avoidance of social unrest and frustration among the masses arising out of certain radical land reforms.
5. The government should reserve a certain minimum amount of crops to combat (a) droughts, (b) floods.

Given the socio-economic conditions prevailing in India today, *i.e.* resources crunch and massive unemployment, the trend must change in favour of small-scale industries based on low-cost, labour-intensive and, as far as possible, on indigenously developed appropriate technology, particularly suited for agro-based industries essential for agricultural mechanisation and rural industrialisation.

In our policy formulation for revitalising the rural economy, it should be borne in mind that we, in fact, are strengthening the industrial base itself. Most of the essential inputs into the modern agricultural production system are, in fact, industrial products such as pesticides, farm machinery, various cultivating equipments, pipelines, electric motors, etc.

Chapter 3: Land Reforms: A Structural Problem in the Formation of the Agro-Industrial Complex; the State of Bihar

The analysis of poverty and polity in India in the post-Independence era since 1947, in order to be consistent and coherent, must take into account (a) the policy of the British colonial plunder on the one hand, and (b) the policies and procedures laid down by the first Prime Minister of India, Jawaharlal Nehru, on the other. This is primarily because the latter, for all practical purposes, is essentially the derivative of the former. "The evidence shows" as wrote Professor J.D. Sethi[9] in 1988, "Nehru did not even have the elementary understanding of land reforms".

A true national development, for any just person, would mean that rural areas and urban centres must share proportionately the benefits accruing after the Independence. Keeping in dark and doom the fate of over 80 per cent of the population in rural areas, therefore, enables one to realise and conclude the meaninglessness of the Independence, as far as their interest is concerned.

To eliminate the gross inequalities between rural areas and urban centres as revealed in chapter 1, we must sincerely attempt to admit and remove the constraints in the task of rural development. The most important of these constraints is, without doubt, the constant failure of the sincere implementation of land reform programmes.

About land reforms, what is being promised ever since the dawn of Independence in 1947 has never really been carried out. To have functional results in land reforms, we must implement sweeping structural changes in our policy formulation and then strictly implement them. The necessity for land reforms are discussed with a case study from the State of Bihar, a State so notorious as to go to any extent to commit atrocities on landless labourers, particularly on the scheduled castes.

Editorials on two different occasions in one of the most prestigious and widely circulated newspapers of India, the *Hindustan Times*, categorically stated the following:

1. The Editorial dated 19-8-1988[10] stated as follows:

   ...For forty years we have been talking of the need for land reforms. Unfortunately, land reform has remained mostly a talking point.

2. The Editorial dated 16-3-1992[11] stated as follows:

> The countries that have succeeded in modernising their economies all began with far-reaching land reform. In our country... The major obstacle in the way of meaningful land reform has been the lack of firm political will at the highest levels.

Chapter 4: Mobilisation of Human Resource

The then President of Tanzania, Julius K. Nyerere[12], said in 1967:

> People, not money, bring about the development of a country. Money, and the wealth it represents, is the result and not the basis of development.

Most of the population of the abundantly populated developing countries such as India dwells predominantly in rural areas. The ignorance and illiteracy of the villagers remain major hindrances in the fuller implementation of many developmental projects. The elimination of illiteracy and the spread of awareness are bound to make the over-population emerge as an asset in development planning, which otherwise remains a liability.

So what is needed is an approach to solve the problems of ignorance and illiteracy, *i.e.* to transform what on surface appears to be a liability into an asset.

When eventually in power in 1947 after the Independence, the leaders of free India virtually set aside the ideas and aspirations of Mahatma Gandhi, the Father of the Nation. What Mahatma Gandhi[13] realised and advocated in 1939, even before Independence, remains to be achieved even after nearly half a century of Independence:

> In the case of the Indian villager an age-old culture is hidden under an encrustment of crudeness. Take away the encrustation, remove his chronic poverty and his illiteracy, and you have the finest specimen of what a cultured, cultivated, free citizen should be.

The attempt in India since Independence has by and large, aimed at solving the problem of population, without first attacking the problem of the people, *i.e.* illiteracy and poverty. Unless we solve these 'problem of the people', the 'population problem' can at best be scratched at the surface only. So, is there 'population problem' or the 'problems of the people'? In the case of India, it is the latter that feeds the former while both, in fact, are the serious problems. It is analyzed in the text that the development planning of the country has got to be linked with the population planning, or else we can have no effective control over population planning.

The World Population Forum for the 21st Century[14], organised in November, 1989, in Amsterdam by the United Nations in cooperation with the Government of Netherlands, recommended that the "population strategy ... must link population programmes to programmes on health, education, housing, and employment, among others."

The policy planners in India appear to have been so thoroughly influenced by the Western concepts of population policies and problems that they virtually ignore the fact that the Western policies simply do not apply to countries like India. Mao tse Tung[15], while refuting "the absurd arguments of Western bourgeois economists like Malthus that increases in food production cannot keep pace with the increases in population, said that the solution to the big population problem "is production".

Chapter 5: Utilisation of Natural Resources

This chapter largely deals with the planning of developmental projects such that the necessary variable inputs are abundantly available in India. These variable inputs are people, land, water, forests, minerals, alternate sources of energy such as solar energy, fisheries, and other aquatic resources, low-cost labour-intensive projects with as much as possible indigenously developed technology, etc. Julius K. Nyerere, former President of Tanzania, had said[16]:

> A poor man does not use money as a weapon. The major means of production are: the land, forests; mineral resources; water; oil and industry...upon which a large

section of population depends for their living, or...which produce essential raw materials.

Chapter 6: Poverty *versus* Polity

In this chapter, the three evolutionary, reformists, and revolutionary approaches to eradicate poverty, or the time taken to initiate the desired social change, are graphically illustrated where the time taken is drawn against social change.

1. *Evolutionary process: The policy of the British colonial plunder*

In the evolutionary process, as the term itself implies, it takes much more time to accomplish very little social change. Some of the accomplishments in the evolutionary process, in fact, are rather the by-products and are "necessary" for the effective plunder of the nation. A slave must not be starved to the point that he dies. This was the policy of the plunder of the British colonialism. The most educated and the most prosperous nation in those times before the British intruded, India had been reduced to one of the most wretched nations of our times.

So the little accomplishments achieved during the British colonial plunder were not meant to be a gift to India: they were in fact, required for effective plunder. There is no shortage of intellectuals[17] in India, not to mention the British and others, who, putting aside all sense of shame and dignity, attempt to point out the "contribution" of the British colonialism in the development of the nation. These so-called intellectuals are nothing but schizophrenic flunkeys suffering from deep inferiority complex, who would do anything that would give them an opportunity to serve their former Anglo-Saxon white-skinned colonial masters.

Of these "accomplishments" during the colonial period, Karl Marx said: "The results must be inappreciable."[18]

The evolutionary approach, or the approach to eradicate poverty to which they were utterly indifferent, was the approach of the British colonial rulers. That is natural, tragic though it is from our point of view in India. The British policies were to plunder the nation, and this was their sole purpose.

Hence during the British period, no changes in rural areas were initiated, and if somewhere they were necessary to effectively plunder the countryside these changes were the undesired by-products. I do not intend to document this plunder by British imperialism any further, as I am primarily concerned in this book with an analysis of poverty *versus* policy in the post-Independence era.

After the end of colonialism in the aftermath of Second World War in 1945, the World Bank and the IMF stood as *saviour* to provide loans and formulate programmes of socio-economic development in the developing countries, *i.e.* countries which were the colonies of the western countries before the Second World War. To maintain *status quo, i.e.* the poor countries continue to remain poor, and the hegemony of western powers remains unchallenged and unquestioned, in the former colonies, the World Bank and IMF prescribed *"solutions"* for the removal of poverty in developing countries. The World Development Report 2000-2001 of the World Bank has prescribed solutions which are nothing but to perpetuate *status quo* (*Detail* in Chapter 6).

Commenting on the World Bank's World Development Report 2000/2001 which is full of prescription for poverty reduction programmes, an Editorial in the *Times of India*[19] stated that:

> But it seems that an organisation which has supported statism in the developing world since its inception—the World Bank is an architect of failed policy is loath to admit it.

2. ***Reformist approach: Opportunism in the post-Independence era since 1947***

In order to get power, the leaders struggling against the British colonialism in India promised many social changes after Independence to remove the grinding poverty of the people.

Jawaharlal Nehru, the first Prime Minister whose 17-year rule was unopposed and unchallenged, clearly implied so by publishing his *Discovery of India* in 1946.

These leaders after the Independence began to indulge in pursuits of their selfish ends, personal glory, and ego-gratification. Corruption, malpractice, and administrative inefficiency became rather "acceptable" traits of our socio-economic institutions, particularly from 1970s to 1990s. Vasant Sathe[20], Cabinet Minister from 1980 to 1989, gleefully admitted in 1988 that "the most popular films and stories are those which caricature and depict men in authority, including judges as being utterly corrupt and without any character."

Under the banner of socialist phraseology, the socio-economic development of the nation in the post-Independence era has been largely based on the Western style urban based modernisation.

While reviewing the book by Sundeep Waslekar, *"South Asian Drama: Travails of Misgovernance"*, Professor K.L. Kamal[21] wrote that:

> "...in all South Asian Countries, governance by personal rule has led to subjugation of the institutions of state to certain individuals and their coteries at different levels. As a result, the ability of the state to govern has grossly declined."

This is what precisely has been hallmark of achievements of more than half-a-century of Independence in this reformist approach aimed at poverty alleviation in all South Asian Countries.

3. *Revolutionary change: The plight of the poor and exploited*

In a revolutionary change towards the development of a nation, particularly the economic development and the eradication of poverty and rural reconstruction, it takes much less time to accomplish the much-needed social changes on a momentous and unprecedented scale.

Although we in India ever since we freed ourselves from the British colonial yoke have always had the officially declared policy of socialism to carry out the task of nation-building, but the fact remains that despite a somewhat decline in the percentage of people below the poverty line, we have absolutely failed to remove poverty. R. Venkataraman, President of India, while addressing the nation on the eve of Independence Day on August 15, 1990 admitted the following[22]:

> One-thirds of the population of India still lives at the subsistence level. The contrast between rich and poor, urbanite and villager, educated and illiterate, urban street-child and the girl-child in particular, suffers from painful disabilities.
>
> The vast mass of our people has remained outside the periphery of our progress.

So before the genuine transformation of rural areas and the removal of poverty is practically possible, we will have to what Bhanu Pratap Singh[23] says, "tear down the existing apparatus of politico-economic feudalism", or else, "a social upheaval cannot be averted." To avoid this inevitable social upheaval, the following policies now to be immediately adopted and implemented:

1. India lives in her villages. A sizeable investment, more than 50 per cent in any case of the investment outlay, should be channelled to the rural areas with sincere implementation of land reforms which have been the backbone of continuing wretchedness in rural areas, for there is no land which the people can call their own.
2. Small-scale and cottage industries should be promoted and protected in rural areas, with a policy that what can be conveniently manufactured in the small-scale sector will not be encouraged in the large-scale sector, with an emphasis on low-cost labour-intensive industries with self-reliant and indigenous technology.

If the above policy alternatives are honestly translated into action, it would certainly remove poverty, which we have not been able to do so in India for nearly the last five decades since Independence in 1947.

The data analysed in this book are mainly from India, South Asian countries, China, and developing countries as a whole. The analysis in the book is applicable to India and South Asian countries in particular, and to the Third World in general.

The nations in the South Asian region, particularly Bangladesh, India, Pakistan, and Sri Lanka, share the same colonial heritage. These nations in South Asia, with somewhat different political systems of governance, have the potential to form a common single unit, economically and culturally. All these nations are the classical victims of:

1. The divide-and-rule policy of the British colonial domination, and
2. The socio-economic neo-colonialisation being practiced by the Western countries after direct political colonialism.

Since we have now been free from colonialism for over four decades, there is a little point in putting the blame on these former colonizers, for it is we ourselves who form our own polity and developmental policies with an urban bias, largely based on western style.

The so-called economic reforms and liberalisation of the economy that began in 1991 have not at all benefited the ordinary citizen among the teeming millions in India, and in any of the South Asian Countries. The two polls carried out by the *Times of India* and the *Economic Times* reflect the opinion of a section of the society who is net users, who overwhelmingly felt that the reforms did not benefit ordinary citizen:

(a) Times of India[24]:

Poll Results:

Question: Has economic liberalisation failed to bridge the rich-poor gap in India?

Yes = 71%, No = 25%, Can't say = 4%

8215 Votes in all

The poll reflects the opinions of Net users who chose to participate, and not necessarily of the general public.

(b) Economic Times[25]:

Poll Results:
Question: Has the Government lived up to the expectations of the common man in the last 10 years of reforms ?
Yes = 21%, No = 75%, Can't say = 4%
This poll reflects the opinion of only those internet users who have chosen to participate.

As far as the interest of the poverty stricken mass are concerned, an eminent journalist of India, Prabhas Joshi[26] wrote about the outcome of the reforms:

> There has been manyfold increase in the capital of the big industrial houses, and all of them have formed partnership with foreign companies. Some percentage of middle class and upper middle class got consumer goods much easily. The role of peasants, the artisans, and the small scale entrepreneurs has been marginalized in these nine years of globalisation.

For the various programmes of rural development for alleviating poverty, the Government of India allotted:

Rs. 7,517 crore in 1999-2000,
Rs. 6,760 crore in 2000-2001,
Rs. 8,197 crore in 2001-2002,
Rs. 7,973 crore in 2002-2003.

As reviewed and analysed in chapter 1, despite plethora of these multi-dimensional programmes of rural development on the one hand, and large sums being spent to implement them on the other, poverty still remains as perennial a problem as ever before. So what is required is a revolutinary approach, a paradigm shift from:

(a) The evolutionary approach of the World Bank and the IMF in the post-Independence era who dictate the policies of the developing countries for removing poverty (or of the British Colonial approach which was in any case, not designed to do any good for the country, except to plunder it effectively), and

(b) The reformist approach being practiced in the last half-a-century since Independence, which has ended up in what K.L. Kamal[27] wrote that:

> ..."in all South Asian Countries, governance by personal rule has led to subjugation of the institutions of state to certain individuals and their coteries at different levels. As a result, the ability of the state to govern has grossly declined."

The political question in this revolutionary approach in alleviating poverty is simple and brutal. To quote K.L. Kamal[28] again:

> "Who will lead the masses, how shall they get organised, how shall they fight the custodians of power who have all the resources at their command, what would be the nature of the institutions and structures that could be envisaged, what institutional restraints would be imposed on the future leaders of the society ?"

NOTES

1. Dhiraj Nayyar, Alleviating Povery: Role of Good Governance and Constitutional Reform, *Economic and Political Weekly*, Mumbai, October 14, 2000, pp. 3739-3742.
2. Julius K. Nyerere, *Freedom and Socialism* (Dar-es-Salaam: Oxford University Press, 1968), p. 235.
3. The World Bank, World Development Report 2000/2001, Oxford University Press, New York, 2000, p. 3.
4. *Ibid.*
5. For an analysis of the pitfalls associated with foreign debt, aid and technology, see the author's other book, *Globalisation*.
6. For programmes of education for skill development, see the author's other book, *Education for Economic Development*.

7. Vasanth Sathe, "Refusal to Think: An Act of Cowardice", in *Yojana Magazine*, New Delhi, 26-1-1988; p. 9.
8. Mahatma Gandhi (1936), *Collected Works*, Vol. 64, Ministry of I&B, New Delhi, 1976; p. 192.
9. J.D. Sethi, *The Times of India*, New Delhi, March 23, 1988; p. 3.
10. Editorial, *The Hindustan Times*, New Delhi, 19-8-1988; p. 11.
11. *The Hindustan Times*, New Delhi, 16-3-1992, p. 13.
12. Julius K. Nyerere, *Freedom and Socialism* (Dar-es-Salaam: The Oxford University Press, 1968); p. 243.
13. Mahatma Gandhi (1939), *Collected Works*, Vol. 68, Ministry of I&B, New Delhi, 1977; p. 266.
14. *The Economic Times*, New Delhi, 3-12-1989; p. 11.
15. Mao tse Tung (1949), *Selected Works*, Vol. IV, Foreign Languages Press, Peking, 1961; pp. 453-59 (22n).
16. Julius K. Nyerere, *Freedom and Socialism* (Dar-es-Salaam: The Oxford University Press, 1968); p. 235.
17. See the author's other book, *Revival of Cultural Identity*.
18. Karl Marx and F. Engels, *On Colonialism* (New York: International Publishers, 1972); p. 83.
19. Editorial, Go Home World Bank, *The Economic Times*, New Delhi, September 29, 2000, p. 12.
20. Vasant Sathe, "Refusal to Think: An Act of Cowardice", in *Yojana Magazine*, 26-1-1988; p. 12.
21. Sundeep Waslekar, "*South Asian Drama: Travails of Misgovernance*", Konark Publishers, Delhi, 1997. Book Review by Professor K.L. Kamal in *Administrative Change*, Jaipur, Vol. XXV, No. 1, July-December, 1997, p. 80.
22. *The Hindustan Times*, New Delhi, 15-8-1990; p. 4.
23. *The Times of India*, New Delhi, 20-3-1989; p. 3.
24. *Times of India*, New Delhi, 16-11-2000, p. 1.
25. *Economic Times*, New Delhi, 24-2-2000, p. 1.
26. प्रभाष जोशी, *जनसत्ता*, 20-5-2000, पृ. 6.

    The above passage is translated from the original in Hindi:

    भूमण्डलीकरण एवं उदारीकरण के 1991 से पिछले 10 वर्षों में 2000 तक निम्नलिखित परिणाम सामने हैं:

    1. बड़े औद्योगिक घराने की सम्पत्तियों में बेतहाशा वृद्धि हुई और इन सबने विदेशी कंपनियों से गठजोड़ कर लिये,
    2. मध्यम वर्ग के कुछ अंश एवं उच्च मध्यम वर्ग को उपभोकता सामग्री आसानी से प्राप्त हुई, एवं
    3. "9 बरस से भूमण्डलीकरण में देश के किसानों, कारीगरों और छोटे-छोटे उद्योग चलाने वालों को किनारे लगा दिया है। औरतों और दलितों की हालत और बिगड़ी है।"

27. Sundeep Waslekar, *South Asian Drama: Travails of Misgovernance*, Konark Publishers, Delhi, 1997. Book Review by Professor K.L. Kamal in *Administrative Change*, Jaipur, Vol. XXV, No. 1, July-December, 1997, p. 80.
28. *Ibid.*, p. 81.

1

The Problem: People Below the Poverty Line

"A violent and bloody revolution is a certainty one day, unless there is a voluntary abdication of riches and the power that riches give and sharing them for common good."

—Mahatma Gandhi[1]*, 1941*

"Despite high economic growth in some parts of Asia, increasing inequality both within and across countries has meant that poverty has remained a persistent problem."

—Raghbendra Jha[2]*, 2002*

Poverty is essentially a relative phenomenon. Poverty has different meanings for different societies. I would therefore first of all like to define what I mean by poverty. To my mind, as far as the present socio-economic milieu of India is concerned, a person is in a state of poverty when he or she is not in a financial position to afford the following five basic requirements of a dignified human life:

- Sufficient food
- Minimum hygienic housing
- Adequate clothing
- Access to primary education
- Essential health care.

With anyone of the five requirements lacking, a person is living in poverty. To afford the basic amenities of life, a person needs certain income. Data on poverty in India, South Asian countries and in developing countries are given below.

Population, GNP, Poverty and HDI Rank in South Asian Countries are given in Table 1.1. Human Development Balance Sheet with respect to poverty and income in countries in South Asia and in developing countries as a whole are given in Fig. 1.1. The World Bank in its World Development Report[3] 2000-2001, a special addition focusing on poverty in the world, summed-up the situation:

> "*Global and regional patterns: income poverty and social indicators*
>
> ...Poor people are active agents in their lives, but are often powerless to influence the social and economic factors that determine their well-being. ... The world has deep poverty amid plenty. Of the world's 6 billion people (in the year 2000) 2.8 billion—almost half—live on less than $2 a day, and 1.2 billion—a fifth—live on less than $1 a day, with 44 per cent living in South Asia. ...
>
> Between 1987 and 1998 the share of the population in developing and transition economies living on less than $1 a day fell from 28 per cent to 24 per cent. This decline is below the rate needed to meet the international development goal of reducing extreme income poverty by half by 2015. ...
>
> Because of population growth, the number of people in poverty hardly changed. But there are large regional variations in performance. East Asia and the Middle East and North Africa have reduced their numbers in poverty—East Asia dramatically so. But in all other regions the number of people living on less than $1 a day has raised. In South Asia, for example, the number of poor people rose over the decade, from 474 million to 522 million, even though the share of people in poverty fell from 45 per cent to 40 per cent. In Latin America and the Caribbean the number of poor people rose by about 20 per cent. ...

> Two regions fared particularly badly. In Europe and Central Asia the number of poverty soared from 1.1 million to 24 million. In Sub-Saharan Africa the number of poor people increased from an already high 217 million to 291 million over the same period, leaving almost half the residents of that continent poor. ... These variations in regional performance are leading to a shift in the geographical distribution of poverty. In 1998 South Asia and Sub-Saharan Africa accounted for around 70 per cent of the population living on less than $1 a day, up 10 percentage points from 1987. ...

In South Asian countries, percentage of population living below the poverty may have come down in the decade from nearly 45 per cent in 1987 to 40 per cent in 1998, but the fact remains that the absolute number of than has increased from 475 million to 522 million. So at the current rate of progress in implementing the poverty alleviation schemes, the poverty would never be out of the face of the South Asia.

Human development index

The HDI (Human Development Index) measures the overall achievements in a country in three basic dimensions of human development—longevity, knowledge and a decent standard of living. It is measured by life expectancy, educational attainment (adult literacy and combined primary, secondary and tertiary enrolment) and adjusted income per capita in Purchasing Power Parity (PPP) US dollars. The HDI is a summary, not a comprehensive measure of human development. (*Human Development Report* 2001, UNDP, Oxford University Press, New York, 2001, p. 14)

Human poverty index

While the HDI measures overall progress in a country in achieving human development, the Human Poverty Index (HPI) reflects the distribution of progress and measures the backlog of deprivations that still exists. The HPI measures deprivation in the same dimensions of basic human development as the HDI. (*Human Development Report* 2001, UNDP, Oxford University Press, New York, 2001, p. 14).

Table 1.1: Population, GNP, Poverty, and HDI Rank in South Asian Countries

Country Region	*Population 1999 millions*	*GNP per capita (US$ & Indian Rs.) 1999*	*Human Dev't Index Rank among 162 countries in 2001*	*Population in poverty 1995*	
				number millions	*as % of total population*
	HDR 2001 (p. 154)	*WDR (p. 275)*	*HDR (p. 141)*	*HDSA 1998 (p. 178)*	
India	992.7	$450 Rs. 20,250	115	328	35
Pakistan	137.6	$470 Rs. 21,150	127	41	29
Bangladesh	134.6	$370 Rs. 16,650	132	55	46
Nepal	18.7	$220 Rs. 9,900	129	10	45
Sri Lanka	22.5	$820 Rs. 36,900	81	4	22
Bhutan	2.0	$510 Rs. 22,950	130	NA	NA
Maldives	0.3	$1160 Rs. 52,200	77	NA	NA
South Asia	1377.6	$440 Rs. 19,800		438	35

Indian Rs. calculated at the exchange rate of Rs. 45 per US$.

Sources: Data extracted from

1. Mahbub ul Haq and Khadija Haq, *Human Development in South Asia 1998*, The Human Development Centre, Karachi, Pakistan (Oxford University Press, Karachi, Pakistan, 1998).
2. United Nations Development Programme, Human Development Report, 2000 (Oxford University Press, New York, 2000).
3. United Nations Development Programme, Human Development Report, 2001 (Oxford University Press, New York, 2001).
4. The World Bank, World Development Report, 2000/2001 (Oxford University Press, New York, 2000).

Table 1.2: Poverty in Various Regions of the World for 2000

Region	*Poverty 1998 People living on less than $1 a day Rs. 45/$*		*Poverty 1987 People living on less than $1 a day Rs. 45/$*	
	Number Millions WDR p. 23	*Per cent*	*Number Millions*	*Per cent*
East Asia & Pacific	278.3	15.3	417.5	26.6
Excluding China	65.1	11.3	114.1	23.9
Europe and Central Asia	24.0	5.1	1.1	0.2
Latin America & Caribbean	78.2	15.6	63.7	15.3
Middle East & North Africa	5.5	1.9	9.3	4.3
South Asia	**522.0**	**40.0**	**474.4**	**44.9**
Sub-Saharan Africa	290.9	46.3	217.2	46.6
Total	1,198.9	24.0	1,183.2	28.3
Excluding China	985.7	26.2	879.8	28.5

...Poor people are active agents in their lives, but are often powerless to influence the social and economic factors that determine their well-being. ... The world has deep poverty amid plenty. Of the world's 6 billion people (in the year 2000) 2.8 billion—almost half—live on less than $2 a day, and 1.2 billion—a fifth—live on less than $1 a day, with 44 per cent living in South Asia. (p. 3)

Source: The World Bank, World Development Report 2000/2001, Oxford University Press, New York, 2000, p 23. The poverty line is $1.08 a day at 1993 PPP)

The data on poverty in various regions of the world for the year 2000 are given in Table 1.2. Per capita income in India during 1980-2000 is given in Table 1.3 as follows:

Table 1.3: Per Capita Income in India

1980-81	*1995-96*	*1999-2000*
Rs. 1830	Rs. 8498	Rs. 10214

Source: Planning Commission, Government of India, New Delhi (*Hindustan Times*, New Delhi/Bhopal, 7-2-2002, p. 1).

It is clear that in the 20 years from 1980 to 2000, there has been a sharp rise in the per capita income, jumping nearly six

I-Poverty and Income

Human Advance	*Human Distress*
South Asia	
GNP grew at an average annual rate of more than 5 per cent, and per capital GNP by 3 per cent during the years 1980-93. Merchandise exports grew at an annual rate of 7 per cent in 1980-93.	The real GDP per capita is still only $1370, which is the lowest compared to any other region in the world. South Asia is home to over 500 million of the poorest people in the world.
India	
The average real GDP per capita (in PPP $) doubled, despite a doubling of the population in the last three decades.	44 per cent of the total population lives in absolute poverty. Nearly one-third of the world's poor live in India.
Pakistan	
In the last two decades, GNP per capita registered an increase of 231 per cent—the highest in South Asia.	36 million people live in absolute poverty. More than half of the cultivable land is in holdings of 50 acres and above, in the hands of big landlords.
Bangladesh	
There are no excessive concentrations of income, wealth, or land compared to other South Asian countries.	52 per cent of the population survives below the absolute poverty line.
Nepal	
Per capita income has more than doubled in the last two decades.	Despite a vigorous growth rate, the current level of per capita income is only $190, the lowest in the South Asia region.
Sri Lanka	
Real GDP per capita (PPP $) more than doubled between 1960 and 1993.	Almost one-third of the population lived in absolute poverty in 1993.
Bhutan	
The Human Development Index for Bhutan has increased by 31 per cent since 1987, which is greater than the increase in any other country in the region.	In real terms (in PPP $), the GDP per capita of $790 is the lowest in the South Asian region.
Maldives	
Per capita income, at $820, was the highest in South Asia.	

Fig. 1.1: Human Development Balance Sheet with Respect to Poverty and Income in Countries in South Asia and in Developing Countries in 1999

Source: Mahbub ul Haq, *Human Development in South Asia, 1997* (Karachi, Pakistan: Oxford University Press, 1997, details summarised from many pages).

II-Poverty and Income

Global Progress	*Global Deprivation*
During 1990-97 real per capita GDP increased at an average annual rate of more than 1 per cent. Real per capital consumption increased at an average annual rate of 2.4 per cent during the same period.	Nearly 1.3 billion people live on less than a dollar a day, and dose to 1 billion cannot meet their basic consumption requirements. The share in global income of the richest fifth of the world's people is 74 times that of the poorest fifth.

Source: UNDP, *Human Development Report 1999*, Oxford University Press, New York, 1999, p. 22.

III-Widening gaps between rich and poor since the early 19th century

World inequalities have been rising steadily for nearly two centuries. An analysis of long-term trends in world income distribution (between countries) shows that the distance between the richest and poorest country was about 3 to 1 in 1820, 11 to 1 in 1913, 35 to 1 in 1950, 44 to 1 in 1973 and 72 to 1 in 1992.

Source: UNDP, *Human Development Report 1999*, Oxford University, Press, New York, 1999, p. 38.

times. However, measurement of poverty differ somewhat depending on who is measuring and what methods are being used. S. Mahendra Dev[4] has summarized the extent of poverty and the methods as given below:

> Trends in poverty can be examined in two ways. One way is to directly estimate poverty ratios from the National Sample Survey (NSS) consumer expenditure. As there is discrepancy in the consumption data between NSS and National Accounts Statistics (NAS), some feel that NSS data is not reliable. Then one can examine poverty situation indirectly by looking at the trends in employment-unemployment and real wages of workers.

Direct Method: Poverty Ratios Using NSS Consumption Data

Has povery worsened in the post-reform period? Since the introduction of economic reforms in 1991, there has been debate whether the reform measures have adverse effects on the social sector and the poor. First, we look at the estimates based on large and thin samples. The estimates for the period 1973-74 to 1998...show that rural poverty declined in the 1980s and it has not declined in the 1990s as compared to 1990-91. In fact, it incresed to above 40 per cent in 1992 and 1994-95. On the other hand, urban poverty declined significantly in the 1990s. Gupta's estimates also show similar trends on rural poverty. However, in 1998 the rural poverty increased to around 45 per cent. His study also indicates that probably growth was not trickling down in the 1990s.

Indirect Method: Employment and Unemployment and Real Wages

We examine here the trends in employment and unemployment and real wages between pre and post-reform periods.

Employment growth: There were apprehensions that economic reforms would lead to reduction in employment growth. The data given in Table 1.4(b) show that the employment growth based on usual status declined during 1983 to 1987-88 period but it showed higher growth during 1987-88 to 1993-94. Current daily status also shows that the growth of employment was over 2 per cent in the 1980s as well as in the early 1990s. Thus there was no sign of decline in the growth of employment at the aggregate level in the post-reform period.

Table 1.4(a): Trends in Poverty: 1993-94 to 1998, All-India

	Datt's Estimates		*S.P. Gupta's Estimates*		*Total*
	Rural	*Urban*	*Rural*	*Urban*	
1993-94	36.66	30.51	37.27	32.36	35.07
1994-95	41.02	33.50	38.03	34.24	36.98
1995-96	37.15	28.04	38.29	30.05	36.08
1997	35.78	29.59	38.46	33.97	37.23

Source: S. Mahendra Dev, Economic Reforms, Poverty, Income Distribution and Employment, *Economic and Political Weekly*, Mumbai, March 4, 2000, p. 824. Estimates based on NSS data on consumer expenditure.

Based on the above methods poverty in India is given in Table 1.4(a) and Table 1.5. Poverty in India is obviously a bit more among the scheduled caste and scheduled tribe population (SC/ST). These SC and ST have traditionally being on the lower side of poverty among the poverty stricken population in India. The NIRD (National Institute of Rural Development), quoting A. Dubey, reported the comparison of SC/ST on the one hand, and general category on the other as shown in Table 1.5. Table 1.5 shows that during 1993-94, poverty among the SC/ST is around 55 per cent while in general category, it is 37 per cent. Further more, poverty is more in illiterate population than in literate population, both in SC/ST and in general categories. Among SC/ST, the poverty stricken literate population is 45 per cent while the illiterate population is 60 per cent. Among general category, the poverty stricken literate population is 30 per cent while the illiterate population is 46 per cent. National Institute of Rural Development, Hyderabad, in its India Rural Development Report[5], 1999 mentioned that:

Table 1.4(b): Rate of Growth of Employment

Period	*Rate of Growth of Employment*	
	Usual Status	*Current Daily Status*
1972-73 to 1977-78	2.73	0.92
1977-78 to 1983	2.17	2.08
1983 to 1987-88	1.54	2.89
1987-88 to 1993-94	2.43	2.52

Source: S. Mahendra Dev, Economic Reforms, Poverty, Income Distribution and Employment, *Economic and Political Weekly*, Mumbai, March 4, 2000, p.825. Planning Commission; based on NSS data on employment and unemployment.

Table 1.5: Poverty (Head Count Ratio) by Social Groups and Literacy

Social Group Year	*SC*	*ST*	*General Category*
1987-88	61.13	66.00	43.65
1993-94	54.65	55.13	36.90
Literacy Group (1993-94)			
Illiterate	60.16	60.00	45.77
Literate	44.57	45.13	29.6

Source: Dubey, A. *Poverty Incidences among the SCs & STs in India*, Paper Presented at IFAD-NABARD Conference on IFAD's Strategic Direction in India, Mussoorie, June 9-10. Quoted in National Institute of Rural Development, India Rural Development Report, 1999, Hyderabad, India, p. 20.

> "Poverty is higher by 15-20 percentage points among the SCs and STs, who are also mainly agricultural labourers, and other labourers, when compared with general category population. Also, the incidence of poverty is much lower among literate."

Regarding the extent of poverty in India, Vasant Sathe, the then Cabinet Minister in the Government of India, had honestly admitted[6]:

> "It is said that the first condition for improvement is the need of a will to look at the facts in their face and willingness to face the truth.
>
> If we deliberately adopt an ostrich-like attitude of shrinking away from facts and burying our heads in the sand of self-delusion, then we can never came to grips with reality nor shall we be able to find a solution to these problems.
>
> The fact is that more than 85 per cent of our people in rural areas and more than 60 per cent of our people in urban areas can be said to be in abject conditions of poverty, if by that it is understood to mean absence of basic requirements of a dignified and hygienic human life."

Having outlined the extent and degree of poverty even after 50 years of Independence, let us now state how Nehru, *"the architect of modern India"* (as stated by the philosopher-President of India, S. Radhakrishnan) perceived the problems of poverty in India before he came to power in 1947. Nehru[7] wrote in his *Discovery of India*, in 1944, that at *"no time did we forget that our main purpose was to raise the whole level of Indian people, psychologically and spiritually and also, of course, politically and economically..."*

So here is the gap between what Nehru promised before Independence and what he and subsequent political masters delivered after Independence, in so far as the poverty of the people is concerned. I would like to let the readers of this book ponder over the fate of the poor of India, especially in rural areas, where poverty and human degradation go together. To quote Nehru[8] further:

"Of our millions how few get any education at all, how many live on the verge of starvation, how many have nothing to look forward to but a clerkship in some office ?... If life opened its gate to them and offered them food and healthy conditions of living and education and opportunities for growth, how many among these millions would be eminent scientists, educationists, technicians, industrialists, writers and artists, helping to build a new India and a new world?"

Twelve years earlier, in 1932, Nehru[9] had also written that it would make more sense "to try to understand the cause of poverty and misery and exploitation, and then try to remove them", rather than "to get angry with a country or with a people, with Britain or the British". Sure enough. This is all right and makes a lot of sense. A positive and constructive approach to a problem is far better than a negative and destructive approach. But the question remains: Did Nehru fail to understand the cause of poverty and misery and exploitation, or did he simply forget about poverty as a problem, virtually becoming indifferent to it?

I am inclined to believe that Nehru deliberately kept the people of India illiterate, poverty-stricken, ignorant, and away from participating in the task of nation-building, so that he could rule, unchallenged and "democratically"; for it was relatively easier to fool and confuse the ignorant and poverty stricken masses. After all, these are the very tools with which the British managed their imperialism and plunder.

Nehru wrote in 1944[10] that "If the spirit of the age demands equality, it must necessarily also demand an economic system which fits in with it and encourages it". Again, we find a gap, a gap that looks ridiculous, between what Nehru thought of Indian poverty and what he actually practiced. Nehru[11] even wrote that "a revolutionary change, both political and economic, is not only needed in India but would appear to be inevitable". The question is the same again: What kind of revolutionary economic and political changes did he deliver to the nation in the post-Independence era? As far as the economic changes are concerned, we noted earlier that the absolute number of the poor may be declining in terms of percentage, but the problem of poverty is far from over because the absolute number of people is so large. Nehru ruled from 1947 till his death in 1964. After Nehru's death, his daughter, Indira Gandhi ruled from 1966 to 1977, and again from 1980 to till her tragic assassination in 1984. Indira Gandhi brought the slogan of *Garibi Hatao, i.e.* remove poverty. But nothing significant at all happened as far as the removal of poverty is concerned during her long 17-years as Prime Minister. After Indira Gandhi, her son, Rajiv Gandhi became Prime Minister from 1984 till 1989. Rajiv Gandhi came with high hopes as a young, energetic, and dynamic and, above all, (as he was widely referred to by the media) as "Mr. Clean". But Rajiv Gandhi too did not deliver anything as far as the plight of people below the poverty line is concerned.

In an effort to eliminate poverty, what is needed in India is intervention by the State to help the poor who are not able to help themselves. President Julius Nyerere of Tanzania and other leaders in that country formulated a policy of socialism and self-reliance, as it appeared in the Arusha Declaration in 1967[12]:

> "That it is the responsibility of the State to intervene actively in the economic life of the nation so as to ensure the well-being of all citizens, and so as to prevent the exploitation of one person by another or one group by another, and so as to prevent the accumulation of wealth to an extent which is inconsistent with the existence of a classless society."

Thus, the only way is to radically alter the current policies to the Gandhian type. Charan Singh[13], in his *Economic Nightmare of India* (1981), said that what we need is to remove the widespread.

> "Urban bias in our planning and administration... What is needed is not a mere amendment but a complete reversal of the present overall policies...a massive investment in

rural areas...is the only answer to the problems of the village... But there can be no radical change in the present policies unless there is a radical change in the power structure that obtains in the country today. "

Now Charan Singh[14] compares rural poverty and the measures to alleviate it in India with those in Tanzania, China, and Vietnam:

"While India unceremoniously discarded Gandhi with such disastrous consequences, other countries, notably China, Vietnam and Tanzania, not only benefited but even succeeded in demonstrating to the rest of the world how the Gandhian type of planning with emphasis on agriculture and small units in industry, was basically right for a dense agrarian economy— at least in the early stages of its development."

After its initial dependence on the Soviet Union, China was quick to free itself from the Soviet apron strings. Despite its uncompromising stand, China, when Mao died, owed no debt to any country and her unemployment problem had practically been solved. Vietnam's achievements are equally spectacular, while Tanzania under President Nyerere has almost become an authentic model for the successful application of the main principles of the Gandhian approach to planning. True, such draconian measures as the Chinese are employing and which the Vietnamese too have adopted in their country, are not open to free societies like India. But economic incentives can be made to take the place of political coercion; statutory protection can be provided to labour-intensive techniques and enterprises, and terms of trade can be turned in favour of an economic policy of its choice even by a government that is not totalitarian".

The World Bank adopted an arbitrary standard to measure the extent and regional concentration of absolute poverty— that a person is in a state of absolute poverty when he or she has an annual income equivalent to U.S. $ 50.00 or less. On this basis, an analysis of all developing countries with populations of more than one million in the early 1970s revealed that[15]:

1. Approximately 85 per cent of poverty is in the rural areas.
2. In all, about 550 million people are suffering from absolute poverty in the rural areas of the developing world in the mid-1970s.
3. About three-fourths of this total is in the developing countries of Asia, with almost two-thirds of the number in only four countries—India, Indonesia, Bangladesh and Pakistan.
4. In contrast, the developing countries of Latin America and the Caribbean account for only about 4 per cent of the population in absolute poverty.
5. Fifty-three countries with per capita incomes above $ 150, taken together, account for only 8 per cent of the absolute poverty in rural areas.

The situation in the developing countries has not changed substantially since 1975. As outlined in the beginning of this Chapter.

V.S. Mahajan[16], while analyzing the three major components of infrastructure transport, power, and education-concluded that larger investment allocation in the urban sector has not been able to affect the combined rural and urban economy of the nation. His arguments that "unless the rural sector is adequately served by infrastructural facilities", where over 80 per cent of the population barely earns its livelihood on mere subsistence level, the industrial growth in the rural sector is lagging behind the urban sector.

Thus what we see today in India, after nearly five decades of *Swaraj* or self-rule, is an urban bias in development and rural neglect and/or plunders as it had been during the British Rule. Mahatma Gandhi said in 1937[17]:

"I know that nowhere in India do villagers earn so much

as one anna per hour in the villages. That is because we have reconciled ourselves to the intense disparity between the haves and have-nots, and because the city people have, perhaps unwittigly, joined in the British exploitation of the village."

While enacting the Constitution on the 26th day of January 1950 more than two years after the Independence August 15, 1947) the founding fathers planned the destiny of India on a socialist pattern of society. We have, however, ended up after thirty years of "planned" development, with what D.P. Sharma[18] calls:

"A bourgeois society where all are traders in goods, power, and influence. With the upper crust of 10 per cent of population sharing almost all the gains of development, it is not surprising that the pattern of production is elite consumption-oriented. Development has had a strong urban bias whereby nearly three-fourths of the population lives in rural areas."

The World Development Report 1990 (a special issue on poverty in the world published by the World Bank), elaborated a dual approach to reduce poverty as follows[19]:

"This Report has emphasised a dual approach to reducing poverty. The element of this two-fold strategy is:

Efficient labour-intensive growth based on appropriate market incentives, physical infrastructure, institutions, and technological innovation.

Adequate provision of social services including primary education, basic health care, and family planning services."

A person remains in the grip of poverty because he has no education/skills to sell to earn an adequate living. Since he is poverty-stricken, he cannot afford to go to school even for primary education. So there is a real paradox here: lack of education and poverty logically feed each other[20]:

1. A person is poor because he has no skill to secure a gainful employment or create his own facility for self-employment, and
2. He has no skills because, being poor, he cannot afford even primary education : he ends up working as a child labour.

In an attempt to remove poverty, the attack therefore has to be on both the fronts. In this book, I am concerned with the programmes of action for the removal of poverty only.

With commitment to the uplift of impoverished masses in rural areas, the author now outlines in the following chapters certain measures, which can be effected in rural areas. These measures include the formation of an agro-industrial complex, based on agricultural mechanisation and rural industrialisation, land reforms, mobilisation of abundant human resources, utilisation of indigenous natural resources, etc. Hence during the last more than 50 years since Independence, with whatever pomp and show we may celebrate 50 years of Independence in 1997, or 50 years of Constitution in 2000, the fact remains as concluded Professor Raghbendra Jha[21] in 2002,

> "Despite high economic growth in some parts of Asia, increasing inequality both within and across countries has meant that poverty has remained a persistent problem."

NOTES

1. Mahatma Gandhi, *Collected Works*, Vol. 75 (New Delhi; Ministry of I&B, 1979), p. 158.
2. Raghbendra Jha, "Combating chronic poverty in Asia", *Economic Times*, New Delhi, 27-5-2002, p. 6.
3. The World Bank, World Development Report 2000/2001, Oxford University Press, New York, 2000, pp. 3, 21-23.
4. S. Mahendra Dev, Economic Reforms, Poverty, Income Distribution and Employment, *Economic and Political Weekly*, Mumbai, March 4, 2000, Estimates based on NSS data on consumer expenditure, pp. 823-25.

5. National Institute of Rural Development, Hyderabad, in its India Rural Development Report, 1999, p. 19.
6. Vasanth Sathe, "Refusal to Think: An Act of Cowardice", in *Yojana*, Vol. 32, Nos. 1 and 2; Jan. 26, 1988, p. 10.
7. Jawaharlal Nehru, *Discovery of India*, (London: Marian Book, 1946); p. 44.
8. *Ibid.*, p. 215.
9. Jawaharlal Nehru, "*India's Quest*" (from Glimpses of World History) (Bombay, Asia pub. 1963), p. 215.
10. Jawaharlal Nehru, *Discovery of India*, London: Marian Book, 1946, p. 535.
11. *Ibid.*, p. 539.
12. Julius K. Nyerere, *Freedom and Socialism* (Dar-es-Salaam; Oxford University Press, 1968) p. 232.
13. Chaudhary Charan Singh, *Economic Nightmare of India* (New Delhi: National Publishers, 1981), pp. 408, 09.
14. *Ibid.*, pp. 394-95.
15. World Bank, "Rural Development Sector Policy" Paper, Washington DC, 1975; p. 19.
16. V.S. Mahajan, *Socialist Pattern in India* (New Delhi; S. Chand and Co., 1974), p. 137.
17. Mahatma Gandhi, *op. cit.*, *Collected Works*, Vol. 66, pp. 124-25, 976.
18. D.P. Sharma, *Times of India*, New Delhi, March 28, 1977, p. 6.
19. World Development Report 1990, The World Bank Washington D.C., 1990; p. 138.
20. For detailed analysis, *see* Sures Chandra Jain, *Education for Socio-Economic Development*, Concept Publishing Co., New Delhi, 2004.
21. Raghbendra Jha, "Combating Chronic poverty in Asia", *Economic Times*, New Delhi, 27-5-2002, p. 6.

2

The Formation of an Agro-Industrial Complex

"In reality, the toiler is the owner of what he produces. ... Who can dispute the fact that the grinding poverty of the masses is due to their having no land which they can call their own?"

—Mahatma Gandhi, 1936[1], 1940[2]

A. THE POLICY ISSUE

In the preceding chapter, while analysing the quantum of poverty and the extent of inequality in India, we noticed that even after 50 years of Independence more than half of India's 800 millions are still below the poverty level. These people have simply been left out both from participation in the process of economic building on the one hand, and the consumption of the fruits of the economic development on the other. Prof. Madhu Dandavate, rightly[3] said after taking over as Finance Minister of India in 1989:

> "Cool and sedate statisticians complacently quote figures of the growth rate of our economy to assess the degree of development. However, the question we must ask ourselves is "Growth—yes; but growth for whom"? In a country where vast sections live below the poverty line...the touchstone of our perspective must be the interest of our economy to sub-serve the interests of the teeming millions."

Taking into consideration the non-linearity between food production and population growth, we face a vast population, and its geometrically accelerating growth. We face a situation in which it is necessary: (1) to feed these people, and (2) to provide them with jobs. The simple answer to the former is 'mass-production' and to the latter 'production out of masses'. This is a very complicated and frustrating job to deal with, and needs a tremendous amount of patience, hard work, and sheer determination. It is necessary to make a clearly defined distinction between feasible and impracticable options, and alternatives available at our disposal. In spite of all this, we have to be energetic, courageous, and have to restore the confidence and enthusiasm. It is a human problem for human welfare, and unfortunately the problem has been created by humanity itself. In my view, what has been missing is a clear conception of the different resources at our disposal and implementation of their optimal use.

I see now a very interesting thing: it is clear that we need 'mass production' and the 'production out of masses'. Hindrances in this 'answer' are probably in the question itself.

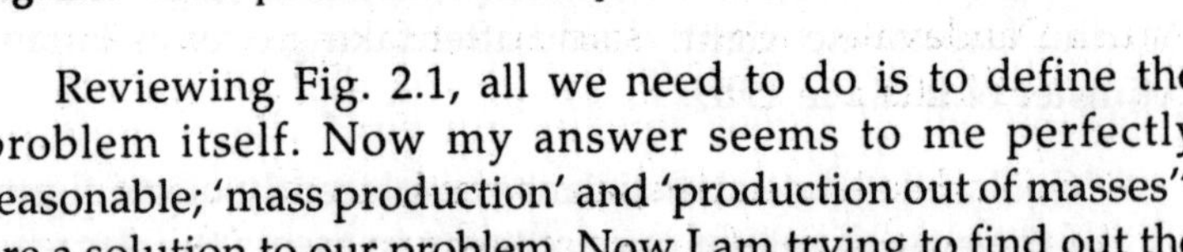

Fig. 2.1: Mass production and the production out of masses

Reviewing Fig. 2.1, all we need to do is to define the problem itself. Now my answer seems to me perfectly reasonable; 'mass production' and 'production out of masses'[4] are a solution to our problem. Now I am trying to find out the right 'question', the solution of which is the 'answer'. One must take into consideration all the possible ambiguities and all the inevitable harassments and frustrations linked with the 'question', most of them having deep roots with the recent colonial era, and some of them being traditional, *i.e.* due to race and culture and environment of the specific regions the developing world. Back to the 'question' again, I would say that the 'question' that fits this clearly defined 'answer' ('mass production' and the 'production out of masses') is the link between agricultural mechanisation with rural industrialisation, which I have named a agro-industrial complex. The dictionary definition of this agro-industrial complex[5] is

> "Agro-industrial complex, a system of branches of the...economy embracing agriculture and industries serving it and bring its produce to the consumer. It includes those supplying the country-side with farm machinery, fertilisers, building materials and other means of the production, and those providing technical services. .."

In my attempt to form an agro-industrial complex for agricultural mechanisation and rural industrialisation, I would now like to outline certain policy imperatives.

Rural Development: Landowners and Tillers

The ownership of land, capitalist form action, the landlords (Zamindars), nationalisation of land, distribution of the produce, and the share of cultivators are all controversial issues, for which there is no single and easy solution, unless a whole sale massacre is carried out. Mahatma Gandhi said[6]:

> "I find great difference between Lenin and Stalin. Lenin's Russia is no more (in 1941). ... I would be the last to minimise the achievement of Russia, but the whole structure is based on force and violence (in 1937)."

Hence I would say that these issues be solved in complete harmony with the consent of each other, as it was in the ancient India, involving the landowners, the tillers, and the State. To this end, we now take into account (a) the ancient India concepts as illustrated by Marx, Gandhi, and Nehru on the one hand, and (b) the twentieth century revolutionary concepts of nationalisation of land as in the case of the USSR and other socialist nations in the communist block.

"In reality the toiler is the owner of what he produces."

—*Mahatma Gandhi*[7], *1936*

"The peasantry wants land and freedom. ... All class-conscious workers want and are fighting for the peasantry to relieve all the land. ..."

—*V.I. Lenin*[8], *1905*

"Those small and extremely ancient communities,...are based on possession in common of the land, on the blending of agriculture and handicrafts,... The land is tilled in common, and the produce divided among the members."

—*Karl Marx*[9], *1868*

"A democratic collectivism need not mean an abolition of private property, but it will mean the public ownership of the basic and major industries. It will mean the cooperative or collective control of the land. In India especially it will be necessary to have, in addition to the big industries, cooperatively controlled small and village industries."

—*Jawaharlal Nehru*[10], *1944*

To remove poverty from rural areas, I feel that agriculture in most part should remain on a small-scale basis, the village being a single unit. In this regard, Mahatma Gandhi said the following, in 1939[11]:

"If they succeed in their own district, there is no reason why the other 249 districts should not be successfully organised. The reasoning is valid if villages were treated as units. It must be confessed that we have not as yet one single village organised in that fashion."

Despite all the high-sounding talk of Green Revolution in India, the fact remains that, except for the Punjab, Haryana and Western UP, both agricultural output and yield remain substantially low in the rest of the country. The yield even in the Punjab, Haryana, and Western UP itself is among the lowest in the world.

It is also a fact, as outlined in Chapter 1, that the absolute number of people below the poverty line is ever-increasing, despite all the high-sounding figures of quantum of production and growth rates. The bulk of this teeming millions below the poverty lines is increasing in rural areas. Whatever be the gains of the Green Revolution, they have, by and large, remained limited to the big agriculturalists only.

We have, therefore, to turn our attention towards the revitalisation of rural economy. This can be achieved through the formation of an agro-industrial complex, involving agricultural mechanisation and rural industrialisation. This is the policy that needs to be implemented for labour-intensive indigenous resources-based small-scale industrial sector in rural areas for rural development.

It is a fact that ever since Independence governmental emphasis has, by and large, been on the development of the industrial sector, financed by the government both in private and public sectors. This was no doubt needed to provide for basic infrastructural facilities for industrial development at large. Now there has been a further swing of the pendulum towards liberalisation of industrial licensing and regulation, picking up further momentum in 1991-92 during the prime ministership of P.V. Narasimha Rao, largely based on capital-intensive and imported high level technology.

Given the socio-economic conditions prevailing in India today, *i.e.* resource crunch and massive employment, the trend must change in favour of small-scale industries based on low-cost, labour-intensive and, as far as possible, on indigenously developed appropriate technology, particularly suited for agro-based industries essential for agricultural mechanisation and rural industrialisation.

In policy formulation for revitalising the rural economy, it should be borne in mind that we are strengthening the industrial base, for most of the inputs into the modern agricultural production system are, in fact, industrial products such as pesticides, farm machinery, various cultivating equipments, electric motors, pipe lines, etc. In this regard, it would be fair if the policy to revitalise the rural economy also includes an industry status for agriculture. This would reverse the hitherto adverse terms of trade in favour of agriculture. Like their industrialists counterparts, the agriculturists should also be equally favoured through governmental decrees, insurance, loans, prices, wages, movement of foodgrains, etc.

B. A MODEL FOR AGRICULTURAL MECHANISATION AND RURAL INDUSTRIALISATION

In the preceding section, we discussed 'mass-production' and 'production-out-of-masses' approach in our formulation of an agro-industrial complex for agricultural mechanisation and rural industrialisation. Now I am attempting to explore all the possibilities and alternatives through a flow chart to link agriculture mechanisation with rural industrialisation. These details are given in Flow Chart 1. This link between agricultural mechanisation and rural industrialisation is based on what Gandhi[12] said "our village face destruction owing to the disappearance of village industries. They can be reinvigorated only by a revival of village industries".

The Flow Chart provides through successive stages of development:

1. Tillers and operators with the highest pay-off in proportion to their output of production, compared to the amount of labour put in.
2. Land-owners as "trustees" with a guaranteed security for their land and a certain minimum pay-off.
3. Incentives to both land-owners and landless peasantry for their land and labour input, respectively.
4. Avoidance of probable social unrest and frustration among the masses arising out of certain radical land reforms.
5. The government should reserve a certain amount of crops to combat

(a) drought, and
(b) floods.

Trustee is defined by Gandhi[13] as follows:

"The property of the State is the property of the people, the Thakore Saheb is their trustee. As trustee, the Thakore Saheb and other members of his family should perform their duty. Through the performance of their duty to the State, they earn the right of taking a certain amount of money from the State. If the ruler and the ruled perform their duty in this way, there is no room for mutual bitterness."

From the Flow Chart 1, it is obvious that the combination of land and labour must provide a great deal of flexibility. This will have two advantages: to produce higher yields per acre and to utilise the labour force maximally.

In the Flow Chart 1, an attempt is made towards devising a model of the sequences of change which would lead to a relatively smooth modernisation of agriculture and the subsequent emergence of industrialisation for rural development. In this model, I am particularly concerned with 'mass-production' on the one hand, and the 'production-out-of-masses' on the other.

The sequences of change are sketched in the model for agricultural mechanisation and rural industrialisation. I have utilised feedback loops, thereby making the model a recurrent cycle of between (a) labour-intensive and labour-saving processes, and (b) agricultural mechanisation and rural industrialisation.

Every attempt is made to avoid the pitfalls faced by the communist countries in the USSR and China, who import billions of dollars worth of grain and corn almost every year. The experience proves that ideological dogma, in the case of agricultural productivity, leads towards doom and disaster.

Let us look, at the Flow Chart, keeping in view the ownership of the land and the distribution of the produce in

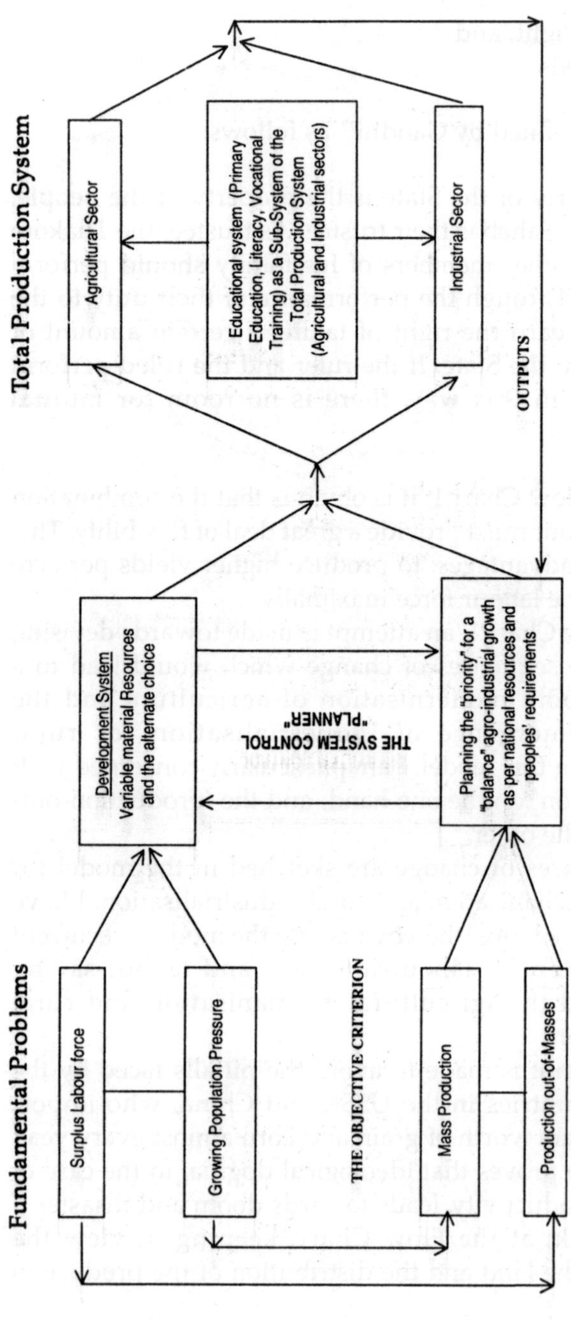

Diagram 1

Decision-making Variables for Agricultural Mechanisation and Rural Industrialisation

Flow Chart 1

Linking Agricultural Mechanisation and Rural Industrialisation for a Generalized Developing Economy

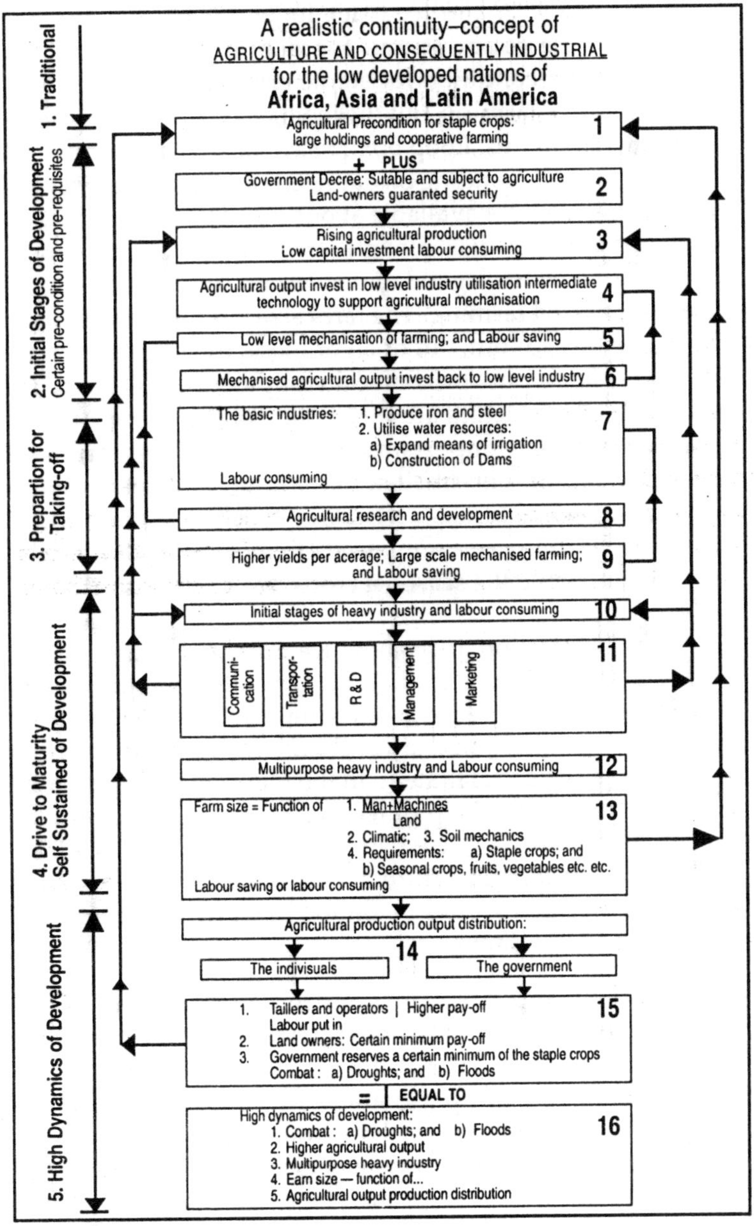

accordance with the views of Gandhi, Marx, Lenin, and Nehru. The Flow Chart is based on certain pre-conditions and pre-requisites (box 1, 2, and feedback loops 13, 14, and 15), and lays down the foundations for economic growth. The agricultural pre-condition for staple crops are large holdings and cooperative farming, but the feedback loop imposes some constraints upon this. Loop box 13 says farm size is a function of many things, depending on our needs and requirements and the various resources available at our disposal. Agriculture by water, buffalo, and disk plough, or by a pair of oxen and a yoke, might well be as good under certain conditions as large-scale mechanised farming. The farm size is a function of many things (see box 13) but, out of all of these, the most important is the ratio (man plus mechanical equipment available to the land area). Other factors are the climate of the region and the soil mechanics. The important factor in deciding the need for large-scale farming or small-scale farming is whether staple crops or the seasonal crops, *i.e.* fruits or vegetables, are required. Staple crop farming certainly needs to replace the yoke and a pair of oxen, while the seasonal crops do not necessarily have to do so.

Now the question arises how we can achieve this goal of uniting these small and traditional agricultural holdings and convert them into large cooperative units to grow staple crops. No doubt, this question is a very sensitive issue, since it has deep roots in the social system. In the Flow Chart, I have tried to explain it with the help of feedback loops. In box 1 and 2, government makes a decree and declares that any available land is subject to agriculture and the resulting production output distribution would be on the basis of feed-back loop box 14 and 15 as follows:

1. Tillers and operators with a highest pay-off in proportion of output production, compared with the amount of labour put-in.
2. Landowners with a guaranteed security for their land and a certain minimum pay-off.
3. Incentives to both, landowners and landless peasantry; for their land and for their labour put-in respectively.

4. Avoidance of otherwise probable social unrest and frustration among the masses arising out of certain radical land reforms.
5. The government should reserve a certain minimum amount of crops to combat (a) Droughts, (b) Floods.

Under this system of distribution (see box 14 and 15), those who own and cultivate the land will be getting much larger proportion of the outcome because of the system: (a) a certain minimum under the government decree (see box 2, 15), and (b) the already highest paid are tillers and operators (same box 2, 14-15). This system has a great advantage. Usually, more fertile areas are under relatively higher tenancy, which discourages the tenants, tillers and operators to invest in the land. With the cooperative distribution of the agricultural output, this lack of incentives would no longer be a constraint on the operators and investors. This distribution of production output avoids small land units which individually cannot be mechanically operated and on which the agricultural machinery cannot be used economically. The law would provide a minimum size of land to be operated for the effective use of machinery.

In this initial stage of development, the role of agriculture is great. In the policy recommendations, and in their planning and in implementation, and in defining the various resources available at our disposal, the role of agriculture becomes a prime-contributing factor. In the Flow Chart, I have shown rising agricultural production (see box 3) after forming the cooperative units (see box 2), without even mechanisation at this stage. It is labour-intensive, and requires low capital investment (box 3).

Next I have invested a very small capital output (box 3) into small-scale industry. At the initial stages, we employ immediate technology and manufacture basic mechanical equipment (see box 4). This mechanises the just-established cooperative farm (box 3), and thus we get low-level mechanised farming (box 5). We see that the two things are interlinked, one being the base for other; the recurrent cycle. The outcome of the agriculture has laid down the foundation for industry, though at a very low level and the cycle thus repeats itself. From the low-level mechanised farming (box 6) we invest back to the low-level industry (box 4), and get the basic industries (box 7). This has been illustrated with the feedback loop (box 6 to box 4 and thus got to box 7, which is the basic industry).

Now I come back to implement these initial stages of development. This goal has been accomplished in different ways in different regions of the world. Cooperative farming is a social phenomenon, corn being the major factor in human life.

Tillers and operators might be able to pursue some other innovative ideas under the system of distribution of the outcome that I mentioned earlier (see box 14, 15), but this kind of ideology can hardly be expected from the labour. Here lies the major difference in the yield per acreage between the collective and State farming systems of the Soviets and Chinese and the farming systems of the Western capitalist States. In the latter, there is a higher yield per acre than in the former, while both of them have almost the same level of mechanisation, equipment, and chemical fertilisers.

The nations of African, Asia, and Latin America have never really been regarded either as developed or underdeveloped. This part of the world has been off the line in modernizing, since it had been under colonial rule. This is indeed a question of getting rid of the existing traditional systems, formulating new strategies, and generating planning and policy recommendations, to keep pace with the rapidly changing times due to technological innovations. The colonial rulers set up the land-systems in their colonies to meet their own selfish ends. Now in the aftermath of colonialism, these existing systems can no longer be tolerated, if we have to increase the food production to keep pace with the growing population. In India the zamindari system was instituted by Lord Cornwallis at the end of the 18th century in an attempt to institute the English landlord system. "There is thus a positive relation between reform and development in the conditions of underdeveloped countries," said Doreen Warriner.[14]

Professor Wittfogel[15] similarly pointed out that the present condition leads us to the realisation that we cannot forever remain on the foundation of the economy of the small peasant. In keeping pace with the progress of the industrialisation of our country, we must make the collectivisation of our agriculture the major constructional task in the rural areas.

Well before the revolution in the Soviet Union in 1917, Marx and Engel's had repeated over again that a Socialist agriculture should be oriented toward large-scale farming. The possibilities of the high yields per acre were well observed for staple crops. After the Revolution in 1917, the lands were nationalised and the collective system of farming came into existence, which, later in the 1930s flourished under Stalin. State farms were introduced. Since 1954, collective farming has largely been replaced by State farming.

The Peoples' Republic of China, on assuming power in 1949, repeated the same story and all the lands were nationalised. In the beginning however, the pressure was toward industry. Seeing no progress in industry, the push was again made to agriculture in 1953.[16] Do we have to be brutal and violent and crush the individual freedom to bring about land reform? Can't we achieve this goal in a peaceful and cooperative way, since it is for the common good of all the members in the society as a whole ?

Land reform, in its political perspective, is too volatile to let the U.N. or any other international agency (Food and Agricultural Organisation or FAO of the United Nations) deal with it. Some might argue that it is a communist way; others might argue it is capitalistic way. I would rather like to see this problem in a different perspective. To feed the hungry and to get rid of hunger is neither communist-style, nor a socialist pattern of society, nor the style of the western capitalist States. It is humanitarian and is the need of the hour in a society, which is rapidly changing due to technological innovations. Of what good is either communism or capitalism to a hungry person? For the nations of Africa, Asia, and Latin America let us keep politics off for a while, as far as the production of corn is concerned.

As I put in the Flow Chart in the beginning (box 1 and 2), the idea of community cooperative farming is a matter of capital accumulation to begin with. Since these nations are 'have-nots', there are two pre-suppositions:

(a) Since the present topic is an "initial stages of development", it gradually develops further and ends up with the "high dynamics of development", we do need to formulate our strategies for further stages of development.
(b) The setting up of small cooperative credit societies is a feasible idea. The banks are usually reluctant to extend credit to individual owners for small farmlands, but for large cooperative single units all the assets of the individuals are accumulated and thus it becomes a corporate entity. The government banks can now extend virtually interest free loans in the beginning to these cooperative societies.

The money thus invested is secured against the assets of these individuals. This system will also break the grip of the individual moneylenders who exploit the peasantry.

After successive stages of development, the redistribution of agricultural outcome (box 15) provides every participating partner with his proper and legitimate share (for details see box 15). (These are also listed in the beginning of this topic).

I may now, well conclude this topic 'initial stages of development' by saying that the "real problems of land reform are those of dynamics, and the analysis should not be cast in a purely static context."[17] This is exactly what I have been outlining in the flow chart, forcing the system to move towards growth with the feedback loops and continuity and eventually towards the dynamic growth (box 16) which is 'High Dynamics of Development', and will be discussed in the following pages.

We have now accomplished the 'initial stages of development' and are heading towards making 'preparation for taking-off stages of development.' The resources at our disposal are tight but dynamic.

Preparation for Taking-Off (While Struggling Hard) Stages of Development

In the previous topic, 'initial stages of development; we were at the threshold of 'the basic industry' (box 7).

At this time, while I am preparing for taking-off, it would be wise to have box 7:

(a) Production of iron and steel, and
(b) Utilisation of the water resources:
   (i) Expand means of irrigation,
   (ii) The construction of dams.

Iron and steel are the major inputs of industrial development. Irrigation facilities are needed to water crops and minimise the effects of droughts. The dam construction will provide water reservoirs for irrigational purposes and control the floods.

These are the major items of expansion for any economy to progress into the self-sustained stages of growth. Next to these are investment on research and development, first for agricultural purposes (box 8).

The outcome of this research has been feedback to the low level mechanisation of farming (box 5), and with this research we hope to get higher yields per acre (box 9).

Now thus I have got the 'higher yields per acre, mechanised farm (box 9) which, together with the already achieved basic industry (box 7), lays down the foundation for the 'initial stages of development' (box 10).

In this strategy of the 'preparation for taking-off', the major constraint is the amount of money we are able to spend on R & D for agriculture (box 8). Before we can take-off, we have to analyse and evaluate:

(a) How much we spend on agricultural research and development (box 8) out of 'basic industry' (box 7), and
(b) How much agricultural research and development affects the yields per acre (box 9).

We now have to measure these R & D effects (out of box 8) quantitatively in their dollar value and qualitatively in their contributions to the yields per acreage.

The various papers[18] indicate that developed nations spend about 2 to 3 per cent of their HNP, while the developing nations spend literally nil (0.05 to 0.5%), on research and development.

The developed nations control about 87.5 per cent of the total GNP[19] of the world with a population of about 34 per cent of the total world; while the developing nations control 12.5 per cent of total GNP of the world, with a population of about 66 per cent of the total world. With this great disparity, the expenditure ratios between the developed and the developing nations range from 40.41 to 1.

The reasons are obvious. The more we spend on research, the more we hope to get higher returns in industry and higher yields in agriculture. This has other effects, too. In the U.S. more than 9 out of 10, and in Europe more than 2 out of 3 persons, are employed outside agriculture, while in South Asia only 1 out of 4 persons' work outside agriculture[20]. This means that the more money we spend on research, the higher is the yield per acreage, and the less people we need to work on farms.

In the 1968 Conference[21] on the Application of Science and Technology to the Development of Asia (CASTASIA), for example, recommended to participating governments a minimum level of 1 per cent GNP on the total national expenditure on R & D as soon as possible, no later than 1980, Similar recommendations were also made to Africa and Latin America.

The important feature regarding R & D is to decide in which fields it is to be carried out, in order to make its possible contribution to improvements in technology. We need a careful coordination between the outcome of research and agricultural and industrial enterprises.

This is the reason that I invested the output of the 'basic industry (box 8) in R & D (Agricultural). Research in private industries being almost nil, it is usually in government-supported laboratories, or in the Universities, or in large multi-purpose corporations who can afford to conduct this research.

The improved varieties of cereals, specially wheat and rice, have developed through a large sophisticated and long process of biological agricultural engineering research.

The Asian Development Bank Survey[22] identifies three major constraints on the research and development:

1. A grave shortage of highly trained manpower.
2. Insufficient funds and other supports from government and mostly a national budget problem, but probably also arising from a low expectancy for research results.
3. Lack of an effective system for coordinating research activities with extension services, sometimes further aggravated by organisational weakness within research centres themselves, which usually results in wasteful overlapping and dilution of efforts.

Regarding the availability of funds, the Survey Report comments:

> "Domestic finance for additional research seems to be the most difficult element in the agricultural development process to obtain, and among the most difficult to find support from external aid sources as well. Yet, it is acknowledged that it is from research that the leverage for development came in the form of new high yielding, responsive varieties, and that it will be from research that the capability of sustaining development will come."

These constraints on research and on funds reflect exactly what I pointed out, that we need good organisation and coordination, careful planning, and communications between the technical researchers and their beneficiaries—the farms and industries. Without a communication media, research is almost meaningless. The communications and management techniques play a vital role in this matter. All these have been discussed in detail in the next topic 'Drive to Maturity—The Self-Sustained Stages of Development'.

Drive to Maturity: The Self-Sustained Dynamic Stages of Development

In the previous topic "Preparation for Taking-off", we accomplished the 'basic industry' (box 7) and 'higher yields per acre (box 9). These together lead us to maturity—the self-sustained dynamic stages of development, the "initial stages of heavy industry" (box 10).

Our objective now is to provide maturity in these initial stages of heavy industry, which have been achieved by the active participation of tillers and operators and government. Now we may consider the conditions of those in the advanced countries; the prime factors being capital accumulations and the highly sophisticated agricultural machinery manufactured in these initial stages of heavy industry (box 10) which are essential for the maturity of our large scale cooperative mechanised farming (box 9).

This provision for maturity is a crucial step. We have to work hard, and even have to bear frustrations due to lack of capital accumulation. But eventually, with our careful planning and coordination[23], we could accomplish it. This is crucial in the sense that industrial growth, being an integral part of the per capita income, provides capital accumulation to invest in mechanical equipment for 'rising agricultural production' (box 3 and feedback loop 11).

Provision of maturity in these initial stages of heavy industry (box 10) leads toward the 'high dynamics of development[24] (box 16) through careful planning and coordination (box 11) and developed means of communications, transportation, R & D (Industrial)[25], management skills, and improved marketing facilities.

Thus we now see that agriculture and industrial processes are cross-links and are becoming an integral process. The achievement of certain preconditions (box 1, 2 and feedback loops box 13-15) however, does not necessarily lead to growth. The initial stages of heavy industry (box 10) and large-scale high-yielding mechanised farming (box 9) need a careful consideration in planning the allocation of resources. Any misallocation might lead to serious consequences, and sometimes complete disaster, during this maturity period.

The four measures, which have been used to improve the conditions of cultivation or farm size, are[26]:

1. Loans and financial assistance,
2. Purchase by the State,
3. Exercise of right of priority purchase, and
4. Expropriations.

With this I may conclude that the role of the government is obviously to increase the size of the holding which I have shown in box 1, to increase food production and to encourage the tillers and operators (for whom I have recommended the maximum pay-off (box 15) to invest as well. I have, thus encouraged the possibility that suitable land would be subject to agriculture and the landowners guaranteed security (box 2 and feed back loop 15):

1. *Box 2*: Government decree, suitable land subjected to agriculture, landowners guaranteed security;
2. *Box 15:*
   (i) Tillers and operators for labour input get the highest pay-off.
   (ii) Landowners: a certain minimum pay-off.
   (iii) Government reserve a certain minimum of the staple crops to combat:
      (a) Droughts, and
      (b) Floods.

In an effort to arrive at maturity and to maintain the self-sustained dynamic stages of development, the increase in the transition from rural to urban areas can never be overlooked. Faster means of communications and better educational facilities[27] constitute the major criteria for this. Providing these facilities, up to a certain extent, in rural areas as well can discourage this. This should then become an integral part of the integrated process of achieving the maturity to maintain the self-sustained dynamic stages of development. Farming ought to be an occupation and looked upon as priority profession. This would need a social change to bring this about, a changing character of the farming population altogether. Co-operative farming, under such governmental intermediately role (box 2), would encourage many people to operate the farmland as their own since operators and tillers are being given maximum pay-off (box 14-15).

Another factor which may hamper the attempt to achieve maturity of sustained dynamic growth would be ineffective management control, poor statistical compilation; marginal analysis of input/output relations to get maximum return on investment; business survey analysis; overall appraisal of agricultural and industrial projects, and careful selection of the farm size. The size of the farm is a function of many variables (box 13), *e.g.* extensive or intensive type of farming, depending upon the ratio of the (man+machines)/land; the climate of the area; soil mechanics, and finally the requirements (1) staple crops or (2) seasonal crops, *i.e.* fruits or vegetables, etc. Before the selection is made, it needs entrepreneur skills and previous statistical compilation to decide with. There should be a higher degree of correlation between the input/output relationships. The input can be figured out:

(a) $\textit{Expenses (Input)} = \dfrac{L + C + M + V}{A}$

where L = Labour input,
C = Capital used,
M = Interest on investment capital,
= Variable cost, *e.g.* gas, oil, repair etc., and
A = Farm area.

(b) Output: the return on the crop is still an unknown factor. This has to be estimated on the previous statistical data and the present market trends.

Owing to uncertainty, no correct correlation can be calculated. In most cases, it has to be estimated on the basis of previous year's outcome and, at a later date, the adjustments made. Greater input does not necessarily mean higher output. In the beginning, the input increases the output; at a later stage, it results in a diminishing marginal way, and further increasing in input may decrease the outcome itself. Thus the amount of input is to be carefully borne in mind and judged by the previous statistical data by experienced and skilled entrepreneurs.

The second thing to be considered is the cost of capital investment. Before we buy any mechanical equipment, we have to consider how long the equipment is going to last, and then the total cost, spread over the entire useful life of the equipment, to figure out yearly depreciation cost analysis purposes.

$$P = \frac{r(r+1)^n}{(r+1)^{n-1}}$$

where

P = Capital investment,
r = Effective rate of interest, and
n = Number of years of useful life of the equipment.

The farm planning and budgeting has four major roles to play in connection with farming in development projects[28]:

1. To reveal the prospective needs of farmers so that a suitable scheme be designed.
2. To determine the best alternative for the organisation of the farm units so as to use the full services and facilities made available after the completion of the development projects.
3. To anticipate farmers' requirements in services and materials other than those provided directly by the projects in order to ensure the success of the project.
4. To furnish basic information for the appraisal of its cost benefit relationship.

The U.N. Development Programme completed a similar programme in Tunisia in 1967-72 regarding farm management, training farm manager's, and farm accountants. For any management and communication system to be effective, we should consider what kind of leadership[29] is needed in the underdeveloped nations to avoid violent and bloody agitation, "from the first day of my reign, I started to warn the nation's leaders to prepare for an evolution from the top before there was a revolution from below".

We have been able to achieve 'multipurpose heavy industry' (box 12) from the 'initial stages of heavy industry' (box 10) by a careful coordination between science, technology and industry, and by management control and communication facilities and the developed entrepreneurial skills (box 11), by feeding these outcomes back to the 'initial stages of heavy industry' (box 10), and then coming on to the 'rising agricultural production' (box 3).

Thus we have now at our disposal a 'multi-purpose heavy industry' (box 12) and 'higher yielding large-scale mechanised farming' (box 9). With these two things, we are now heading forward towards the 'high dynamics of development (box 16), which is a combination of several variables and achievements.

High Dynamics of Development

By now we have achieved and accomplished the following:

1. Combat:
   (a) Droughts .... box 7, 15
   (b) Floods
2. Higher agricultural output .... box 9 (box 16)
3. Multipurpose heavy industry .... box 12
4. Farm size = function of .... box 13
5. Agricultural output distribution .... box 14-15

The 'high dynamics of development' (see box 16) is a combination of all the above-mentioned achievements. We now have to formulate a central command to coordinate all these achievements to enable us to make optimal use and take into account the various constraints on each of the achievements and make this system a 'dynamic system' that repeats itself in a controlled cycle.

If we understand the cross-link between agriculture and the industry in a realistic way (discussed earlier with the help of a flow chart through various stages of development) and make up our minds that agriculture is the foundation for economic development of the present low developed nations,

I hope this problem can be solved and the present misery and gloomy situation spreading all around India and South Asia can be averted in favour of happiness and prosperity in the life of the people. The other nations in Asia, Africa, and Latin America, the so-called "The Third World", too may take a lesson. Now let us have a look at Table 2.1, giving details of average index of agriculture and food production in developing and developed countries.

Table 2.1: Agriculture and Food

	Average index of food production per capita 91979-81 = 100 1987-89
Low-income economies	116 w
China and India	122 w
Other low-income	103 w
Middle-income economies	101 w
Lower-middle-income	99 w
Upper-middle-income	103 w
Low-and middle-income	112 w
Sub-Saharan Africa	95 w
East Asia	123 w
South Asia	112 w
Europe, M. East, & N. Africa	99 w
Latin Amjerica & Caribbean	105 w
Severely indebted	105 w
High-income economies	99 w
OECD members	98 w
+Other	123 w
Other economies	110 w
World	110 w
Oil exporters (excl. USSR)	101 w

- Low-income economies are those with a GNP per capital of $ 580 or less in 1989.
- Middle-income economies are those with a GNP per capita of more than $ 580 but less than $ 6,000 in 1989. A further division, at GNP per capita of $ 2,335 in 1989, is made between lower-middle-income and upper-middle-income economies.
- High-income economies are those with a GNP per capita of $ 6,000 or more in 1989.

Source: Table condensed from World Development Report, The World Bank, Washington, DC, 1991, pp. 210-11, p. X also.

Table 2.1 shows, among other thing, the differences between per capita food production for the developed and the underdeveloped regions. The reasons for this are obvious. The nations, having gained Independence from their former colonial rulers, made a common mistake by pushing industry, while ignoring agriculture. The fact is that, for a economic development, large holdings (box 1) are almost a precondition. In this sense, agriculture is the pre-requisite for industrial development. Industry has to mechanise agriculture. The cycle repeats itself, and the result is economic prosperity.

Under the present situation and without meeting these preconditions (box 1, 2) and feedback loop (box 13, 14, 15), I fear that the developing nations would never be able to achieve economic freedom. The developing nations will never be able to offset their grain import bills, unless there is increased output from their own agriculture. If these measures are taken (box 1, 2 and feedback loop box 13, 14 and 15). I am sure that the gloomy situation, which I have already described in Chapter 1, will be averted and the low developed[30] nations will then be able to get rid of the necessity of big loans from developed nations, the payment of interests on them so often hamper their development projects.

This cooperation among operators, tillers, landowners, and the intermediary government role will certainly provide sound economic basis for the

(a) Supply of food...(thus reducing the import bills)
(b) Consumer goods...(from the industry which will eventually emerge out of rising agricultural production (see box 12).

India, whose population is more than that of combined Africa and Latin America, is beginning to push as a result of her direct heavy industry largely being supported by foreign powers. Thus the planners of the nation envisioned industrial growth without being agriculturally self-dependent. The agricultural "responsibility" was that of the U.S. and the P.L. 480, which have proved a nightmare. During the 1965 war with

Pakistan, the late Prime Minister of India, Lal Bahadur Shastri realised its hollowness when America stopped all the shipments to India. Prime Minister Shastri declared we would eat only one meal. Thus the nation, at least for a while, realised the nightmare of this sad state. Unfortunately, Shastri did not live long enough and died shortly after in Tashkand, USSR, while negotiating, after the 1965 war, Pakistan with President Ayyub Khan. The Chinese, in 1949, made the same mistake in pushing industry first and ignoring agriculture. Soon the mistake was realised and, in 1953, the push was again made back to agriculture.

I would like to conclude by stating that these stages of development would certainly result in the elimination of poverty from the nations of Africa, Asia, and Latin America. I hope that we will no longer have to see human dignity in the same way it has been depicted in a Bolivian[31] newspaper where the following advertisement appeared:

> "*For Sale:* High land farm, thirty minutes drive from La Paz, five hundred sheep, plenty of water, twenty peon"

Why are these nations in Africa, Asia, and Latin America poverty-stricken? Why do the people in these regions struggle only for food? Do these people merely want to 'live to eat'? When in the life of these people will there be choice not between 'live to eat' but 'eat to live'? Will it be before it is too late? Can a hungry person ever be expected to contribute his/her proper share to the 'great drama of unknown'?

Dr. Paul R. Ehrlich[32] points that, while we are reading these words, five people, mostly children, have died of starvation.

When we look at these poor nations, especially in Africa and in Asia, we see that they carry a long history of oppression and exploitation by their colonial rulers and have been consistently discouraged from industrializing and encouraged to produce handicrafts, which put them in a never-growing financial status. Let us look at Table 2.2, which gives us the years of the beginning of industrialisation in various countries.

Table 2.2: Beginning of Industrialisation in Countries

Country	*Dates of Beginning of Industrialisation*
U.K.	1783-1802
France	1830-1860
Belgium	1833-1860
U.S.	1843-1860
Germany	1850-1873
Sweden	1868-1890
Japan	1878-1900

Source: Data condensed from Agricultural Organisation and Development by Xavier Flores, International Labour Office, Geneva, 1971, p. 29, Table 2.

From Table 2.2, we see that UK was going through industrialisation in the last quarter of the 18th century, while Japan and U.S. in the first half of the 19th century. Thus, while the so-called 'haves' of today had actually begun to industrialise two centuries ago, the 'have-nots' of today were being oppressed and exploited.

I reiterate 'mass production' and the 'production-out-of-masses' can will be achieved with self-reliance based on land reforms, mobilisation of human resources and utilisation of natural resources. (All these issues are discussed in Chapters 3, 4 and 5 of this book.)

Before we embark upon generating any policy-planning and decision-making system, we may consider the following three distinct ways:

1. Stating the long-term goals,
2. Short-term planning, and
3. Near-term programming and/or immediate requirements.

A close coordination and integration among these three aspects of policy planning is inevitably required and, of course, policy planning should be reviewed within these three broad forms for successful implementation. This is particularly necessary when we face the lack of resources, financially and in material inputs, management machinery, and lack of alternate choices to execute the programme(s).

A substantial increase in agricultural production and industrial growth (to mechanise agriculture) are vital aspects of development strategies and provide the necessary infrastructure to sustain overall economic growth. The decision of policy planning regarding where to start, is largely a matter of priority and is a function of generally interdependent factors such as the availability of human resources, general level of education and occupational/professional training of the labour force, stock and mining facilities for the natural resources, availability of land and capital, and the need for savings and investments (in other than agro-industrial sectors of economy)—the supplementary sectors such as literacy and vocational education within the industry, and so on. The magnitude of these factors requires that planning should be a matter of priority (planned in the long-run with short-term perspectives and near-term programmes) and availability of resources, for a balanced economic growth. A "balanced" growth means that all significant sectors of the economy (such as education, urbanisation, transportation, communications, health and community development programmes, non-formal education for promoting literacy and accelerating rural development, vocational education within the industry, etc.), must be considered simultaneously as relevant. Thus by a balanced growth, what we need as planners is a coordination between various issues on the one hand, and alternatives to solve those issues on the other. Simultaneously, there should be linkage of agricultural mechanisation with rural industrialisation.

The main contents have been sketched out in Diagram 1. The diagram is self-explanatory. Here I will only point out certain technical details to explain the diagram and the underlying assumptions. The diagram is divided into the following main categories:

1. The fundamental problem,
2. The objective criterion,

3. The system-control, "Planner",
4. The "balance" and/or "Priority",
5. The production division,
6. The composition of the outcome, and
7. Feedback—the underlying assumptions.

The Fundamental Problem (box 1 in the diagram)

The two main problems facing the India planners are the surplus labour force (largely rural), and the growing population pressure. The growing population pressure is such that it virtually offsets whatever economic growth has been achieved. The surplus labour comprises "many manifestations of underutilisation of human resources, open unemployment, under-employment, disguised unemployment, or mal-employment. Moreover, it is difficult to make a distinction between poverty and underutilisation. The statistics on various kinds of underutilisation are very inadequate. Nevertheless, the problem of underutilisation is obvious: manpower, the most abundant of all resources in the developing countries, is grossly underutilised, even though it is capable of almost limitless development.

The Objective Criterion (box 2 in the diagram)

To tackle the growing population pressure and the surplus labour force, I devised an inter- and cross-link mechanism such that there should be linearity between growing population and the production of consumer goods, particularly food production. A similar mechanism shows (box 1) that, if there is surplus labour force, production-out-of-masses techniques (labour-intensive devices), should be employed in the production division (box 5). These two objectives must be fulfilled in any successful and realistic agro-industrial planning.

The System-Control—"Planner" (box 3 in the diagram)

This diagram, the inter-sectorial relationship between agriculture and industrialisation in an underdeveloped agrarian economy with surplus labour force and mounting population pressure, is a two-sector model with two main objective-criteria, as mentioned above. Having considered the dimension and pervasiveness of the fundamental problems and objectives as a criterion, the "planner" should formulate the strategy (box 3 System Control), taking into account the various variable material resources and alternate choices. Then the planner should try to integrate and coordinate these national resources with peoples' needs and requirements on the "priority" basis for a "balanced" national socio-economic growth. The immediate needs and requirements of the people, are, in my judgment, food, clothing, literacy and shelter.

"Balance" and/or "Priority", in the Production Division (boxes 4, 5 in the diagram)

The Diagram, though inter- and cross-linked, provides a basis for planning the "priority" for "balanced" agro-industrial development, once the national resources and the peoples' requirements are fully realised. What is therefore needed is to formulate a "balanced" investment policy in both of the main sectors such that there is no "imbalance" in the composition of the outcome (box 6 in the diagram)—either food deficit or industrial sluggishness. The main supplementary sector, during the initial stages of economic growth, should be in relation to agro-industrial growth:

1. Functional literacy, and
2. Vocational education within the industry.

This "balanced" strategy for agro-industrial growth should be able to:

1. Avert the problem of food-deficit— Agricultural sector;
2. Mobilise the surplus labour force—Production-out-of-Masses;
3. Provide consumer goods—The Industrial Sector.

Composition of the Outcome and the Feedback (boxes 6, 3)

The balanced strategy to produce the required composition of outcome (box 6) from the Production Division (box 5) depends upon how much we invest in specific sectors: (1) agriculture; (2) industrial; (3) supplementary. One of the main strategies for formulating the balanced investment policy to get the required composition of outcome (box 6) as per needs and requirements (box 3, Feedback) is to invest in the sectors where the ratio *k*, marginal (capital/output), is lower. In an under-developed agrarian economy (where over 80 per cent of the population lives on agricultural production, largely at subsistence level), with surplus labour force and food deficit, and where yields are substantially lower in agriculture than elsewhere, any new unit of capital investment in agriculture will bring more return and therefore a higher rate of saving in the economy (and thus a higher rate of capital accumulation) than the same amount of capital investment in the industrial sector. In other words, under the present economic conditions of India, the potential yield from extra resources in the un-mechanised agricultural sector is higher than in industrial sector. It would, now, be pertinent to mention that W. Arthur Lesis' (Labour Surplus Model) argument of shifting surplus agricultural labour for producing capital does not apply, at least in the case of India.

The Underlying Assumptions

The functional validity of planning the entire system control and feedback between programming the individual clauses and then their successful implementation (box 3) depends upon three main assumptions:

1. Surplus Labour force (under- and unemployed, primarily the rural);
2. Food-deficit (un-mechanised agricultural sector, low agricultural yields);

3. The major reforms in the existing socio-institutional structure and attitudinal thought pattern of the conservative society— the necessary land reforms.

The first two assumptions, the two main elements in the present Indian economic scenario, are quite obvious and require no proof. The third assumption may seem rather hypothetical at first, but in fact this is quite a necessary requirement for the successful formulation and implementation of the policy programmes; otherwise, the existing economic dismay and stagnation (food-shortage and industrial slow-down) will continue. The fundamental validity of this diagram depends upon in the "truth" and "possibilities" underlying in the third assumption.

The magnitude, dimension, and the pervasiveness of the contents in the third assumption, however, are easy to recognise, yet difficult to define. A viable cooperation and coordination among sociologists, economists, and the policy planners, therefore, is a realistic necessity.

C. AGRICULTURE AND/OR INDUSTRY

"Villages have remained untouched by progress and many of them still live in Biblical times, while the urban centres are booming.

The smart ones among the rural people have also gravitated to towns and cities where the gravy is, even though living conditions for them are horrendous."

—G.P. Jain[33], 1989

In terms of growth rates and per capita total production in agriculture and industry, developing countries are far behind the developed nations. After 40 years of Independence in India, the share of agriculture and industry in the GDP is approximately 30 per cent each; the remaining 40 per cent belongs to the service sector. Industrial productivity and agricultural yield in India are still amongst the lowest in the world.

Table 2.3 gives GDP by sector growth rates for Agriculture, Industry and Services for various country groups in the world. Table 2.4 gives structure of production for Agriculture and Industry, as percentage of GDP, for various country groups in the World.

To mobilise virtually a stagnant economy with surplus labour force, largely rural (80 per cent) and illiterate (70 per cent unable to read and write) is indeed a huge task facing the new nations emerging in the aftermath of the disintegration of the Colonial-systems after the Second World War. The problem becomes even worse if multiplied by food-deficit with no significant industrial growth. In this Chapter I will attempt to explore the vital problems of the development and modernisation of backward societies. I will mainly focus on formulating the strategies for agro-industrial growth, and review the policy-planning system of India after Independence (1947). I will try to point out where the mistakes (in my judgment) have been and what the future course of action should be to avert the constantly continuing problem of food-shortage and industrial stagnation—a serious imbalance between agro-industrial growth since the Independence.

The Indian system of planned economic growth (through Five-Year Plans) is, by and large, a replica of the Soviet economic system. Nehru, the principal architect of the "modern" Indian image and identity, fighter against Colonialism, a firm believer in human "dignity" and "freedom" and other such ideals, was deliberately against, in his policy-planning for the nation-building, what Professor Moore[34] called the brutality of Communist modernisation:

In response to such an assessment, the Western liberal observer almost automatically responds that even if Indian agrarian policy, indeed as her economic policy as a whole, has rather been long on talk and quite shorter on accomplishments, at least there has not been the brutality of communist modernisation.

The question arises: what does the "brutality of Communist modernisation" or the "ideals of democratic principles" mean to a person who has no opportunity to discover his potential (no access to education even at primary level) and who suffers from serious malnutrition and may even die of starvation) ? Nehru was highly impressed by the planned Soviet economic development, particularly in agriculture and industrialisation. In the Soviet system of planning, industrialisation envisages a comprehensive industrial structure based upon heavy industry. According to the author of *Asian Drama*, the Swiss Nobel Laureate Professor Gunnar Myrdal[35]:

> "This pattern is now often accepted as a natural one for any large under-developed country to imitate. It has specifically influenced the thinking of many South Asian intellectual leaders who were at one time Communists or Socialists (with Communist leanings) among them Jawaharlal Nehru... Almost all traditionalist intellectuals in the region have been decisively influenced by the Communist doctrine of planned and directed industrialisation as a technique for engendering development. To them all, Soviet successes in planning convey lessons they would like to apply when charting their own course of development."

In my judgment, this conclusion of Professor Myrdal exactly indicates the missing link in India's formulation of socio-economic development strategies and in her policy planning system, particularly in agriculture and industrialisation. The missing link is that the Soviet heavy industrialisation programme was substantially fed on agriculture surplus and exploitation.[36] Thus it would appear that there was a wrong approach to begin with in the Indian planning to adopt the big pushes industrialisation without a parallel support from agricultural sector. What makes the matter worse is that Nehru did not follow any radical land reform programme. In contrast, the land reforms in the Soviet Union came as a result of the Russian Revolution of 1917 and the Soviet people were made to pursue the State policies. What makes the matter even more disappointing and frustrating is the fact that, as concluded by W. Nicholls[37]:

Table 2.3: GDP by Sector Growth Rates (average annual percentage change)

S. No.		*Agriculture*			*Industry*			*Services*		
		1965-73	*1973-80*	*1980-89*	*1965-73*	*1973-80*	*1980-89*	*1965-73*	*1973-80*	*1980-89*
(1)	*(2)*	*(3)*	*(4)*	*(5)*	*(6)*	*(7)*	*(8)*	*(9)*	*(10)*	*(11)*
1.	Low-and middle-income economies	3.1	2.5	3.3	8.3	4.9	4.5	7.3	6.4	3.5
2.	Low-income economies	2.9	2.1	4.0	8.8	6.6	8.6	5.8	5.5	6.2
3.	Middle-income economies	3.3	3.0	2.6	8.1	4.4	3.0	7.7	6.6	2.8
4.	Severely indebted middle-income economies	2.8	3.2	2.1	7.4	5.9	1.5	7.3	5.9	1.9
5.	Sub-Saharan Africa	2.4	1.1	2.0	10.4	4.3	0.7	3.4	4.2	2.3
6.	East Asia	3.2	2.5	5.2	12.4	9.4	10.4	9.8	7.2	7.7
7.	South Asia	3.1	2.2	2.9	3.9	5.5	6.7	4.0	5.3	6.3
8.	Europe, Middle East, and North Africa	–	–	2.9	–	–	3.1	–	–	2.7
9.	Latin America and the Caribbean	2.8	3.3	1.9	7.5	5.4	1.6	7.5	5.8	1.6
10.	High-income economies	–	–2.3	1.5	10.9	1.9	2.1	12.6	0.7	3.1

(Contd.)

(Contd.)

S. No.		*Agriculture*			*Industry*			*Services*		
		1965-73	*1973-80*	*1980-89*	*1965-73*	*1973-80*	*1980-89*	*1965-73*	*1973-80*	*1980-89*
(1)	*(2)*	*(3)*	*(4)*	*(5)*	*(6)*	*(7)*	*(8)*	*(9)*	*(10)*	*(11)*
11.	OECD members	–	–2.5	1.3	10.9	1.7	2.2	12.6	0.6	3.1
12.	Other economies	–	–	–	–	–	–	–	–	–
13.	World	2.4	0.2	2.6	10.6	2.3	2.4	12.1	1.2	3.2
14.	Oil exporters (excluding USSR)	4.1	2.7	5.4	9.8	1.4	–1.1	7.9	9.8	1.9

- Low-income economies are those with a GNP per capita of $ 580 or less since 1989.
- Middle-income economies are those with a GNP per capita of more than $ 580 but less than $ 6,000 in 1989. A further division, at GNP per capita of $ 2,335 in 1989, is made between lower-middle-income and upper-middle-income economies.
- High-income economies are those with a GNP per capita of $ 6,000 or more in 1989.

Source: World Development Report, 1991, The World Bank, Washington, DC, 1991, pp. 187, X.

Table 2.4: Structure of Production (percentage of GDP)

S. No.		1965		1973		1980		1985		1987		1988		1989a	
		Agri-culture	*Indus-try*	*Agri-culture*	*Indus-try*	*Agri-culture*	*Indus-try*	*Agri-culture*	*Indus-try*	*Agri-culture*	*Indus-try*	*Agri-culture*	*Indus-try*	*Agri-culture*	*Indus-try*
(1)	(2)	(3)	(4)	(5)	(6)	(7)	(8)	(9)	(10)	(11)	(12)	(13)	(14)	(15)	(16)
1.	Low-and middle-income economies	31	31	24	34	19	38	19	36	18	36	18	36	19	38
2.	Low-income economies	44	28	38	32	33	37	33	33	31	33	31	34	32	37
3.	Middle-income economies	19	34	15	35	12	39	12	37	13	37	12	37	12	36
4.	Severely indebted middle-income economies	17	33	14	33	11	37	11	36	11	36	11	36	–	–
5.	Sub-Saharan Africa	41	20	31	25	28	32	33	26	30	25	31	24	32	27
6.	East Asia	42	35	35	40	29	44	27	41	25	42	24	43	24	44
7.	South Asia	44	21	43	19	35	22	31	24	29	24	30	23	32	26
8.	Europe, Middle East, and North Africa	–	–	16	–	13	41	14	37	15	37	15	35	15	–
9.	Latin America and the Caribbean	16	33	12	33	10	37	10	37	10	37	10	37	–	–

(Contd.)

(Contd.)

(1)	(2)	(3)	(4)	(5)	(6)	(7)	(8)	(9)	(10)	(11)	(12)	(13)	(14)	(15)	(16)
10.	High-income economies	5	42	4	37	3	37	3	34	3	31	–	–	–	–
11.	OECD members	5	43	4	37	3	36	3	34	3	31	–	–	–	–
12.	Other economies	–	–	–	–	–	–	–	–	–	–	–	–	–	–
13.	World	10	40	8	37	7	37	6	35	5	32	–	–	–	–
14.	Oil exporters (excluding USSR)	–	–	13	48	10	54	14	38	13	35	14	35	14	35

- Low-income economies are those with a GNP per capita of $ 580 or less since 1989.
- Middle-income economies are those with a GNP per capita of more than $ 580 but less than $ 6,000 in 1989. A further division, at GNP per capita of $ 2,335 in 1989, is made between lower-middle-income and upper-middle-income economies.
- High-income economies are those with a GNP per capita of $ 6,000 or more in 1989.

Source: World Development Report 1991, The World Bank, 1991, Washington, DC, pp. 186, X.

> "India has, despite relative scarcity of land, tremendous opportunities for increasing yields per acreage by modest capital expenditures on land- saving techniques and a fuller utilisation of her abundant agricultural labour force."

England and Western Europe, for example, also had food surplus because the agricultural revolution had already provided the domestic food surplus and they were able to initiate the industrial revolution. So is the case with Japan and China[38]:

> "Japan, with a minimum of structural disturbance to an agricultural economy characterised by land shortage and a labour surplus, established policies which created a food surplus and became the basis for its spectacular economic development..."

In China, after the Revolution in 1949, though initially the push was towards the Soviet model of industrialisation, later in the 1950s, she too realised the "importance of achieving an agricultural surplus and adapting the Soviet model to its own special conditions,"[39] India, has, ever since Independence, been an importer of food grain, and/or oil seeds without a parallel import-substitution, and the trend continues even after nearly five decades of Independence.

Now one may ask what is and what was the criterion of the Indian economic policy planning for a push towards industrialisation despite its slow growth rate, and not much attention was given to agriculture despite its greater potentialities for growth. Myrdal[40] seems right when he concluded that:

> "When the intellectually elite (South Asian) say their countries are under-developed, they mean in the first instance, that they have too little industry. For instance, India's Second Five Year Plan stated bluntly "rapid industrialisation and the diversification of the economy is the core of development". The spokesmen for the South Asian countries frequently use the terms "pre-industrial"

or "under-industrialised" as synonymous for "poor" or "underdeveloped".

To provide a linear relation between food production and growing population pressure, however, is a very complicated phenomenon to deal with. Let us look again at Diagram 2.2 (of the preceding Section B), which gives us decision-making variables for agricultural mechanisation and rural industrialisation. No single equation or sets of equations may contain all the variable material resources to provide the required composition of the output (box 6 in the diagram) according to national needs and requirements. What is therefore required is that a clearly defined distinction be made among the goals and objectives based upon feasible realities. Increased food production in the composition of agro-industrial output (box 6) has to be an indispensable objective. This is necessary to keep pace with the compounded growth of human population (box 1). According to FAO[41] estimate, there should be

> "A two-fold increase in food output by 1980 and more than three-fold by the end of the century. Unless some drastic measures are taken to remove the non-linearity between food production and taken to remove the non-linearity between food production and geometric growth of population, "we could only delay, not avert, the ultimate collapse."

One of the main strategies for formulating the "balanced" investment policy (box 4) is to invest in the sectors where the ratio *K*, marginal (capital/output), is lower. In an underdeveloped agrarian economy, with surplus labour force and food shortage, and where yields are substantially lower in agriculture than elsewhere, any new unit of capital investment in agricultural sector will bring more return (and therefore the higher rate of capital accumulation) than the same amount of investment in the industrial sector. In other words, the potential yields from extra resources in an un-mechanised agricultural sector are higher than in the industrial sector. What is therefore needed is to formulate the "balanced" investment policy in both of the main sectors, such that there is no "imbalance" in the composition of the outcome (box 6), in either food deficit or industrial slow-down.

The disparity between investments in agriculture and allied sectors and in industry and minerals is quite wide in India. With the exception of the First Five Year Plan, the investments in industry have been twice as much as those in the Second- and Third Plan periods and almost one and a half in the Fourth Plan.

Considering the whole issue of capital investment in agriculture and industrialisation in terms of output differentials, and where marginal yield of agricultural outlay is much higher than in any other sector, the agricultural sector demands a bigger share of the total investment. In contrast to India, since early 1950s:

> "China puts more into agriculture, which has frequently turned out to be a sector where relatively large returns follow from a unit of new capital—at least in underdeveloped areas where yields are very low initially[42] ...
>
> I am convinced—on the basis of my brief tour of villages in both South and North India in 1958- that the return on modest investments in such key limitation factors as fertilisers, improved seeds, insecticides and pesticides, and better cultivation and irrigation practices would be extremely high."[43]

However, since 1958, the same trend continuous and the investments in the industrial sector have been substantially higher compared to the agricultural sector. Here it would be pertinent to make a differentiation between village and small industries and large-scale industrialisation. There are several arguments for investment in small-scale industrial sector in favour of agricultural sector. First, India spends just about 10 per cent of total industrial investments on the small-scale industry. Second, if India increases investment in the small-scale industrial sector, unless supported by external economic inputs (food surplus and agricultural base to cater to the mass needs), it has a combined effect, which is less than the sum of all the investments in industrial projects. This is partly because of the constraints of the small-scale enterprise to offer any substantial new employment opportunities. And, therefore, as far as the efficiency of the favourable spread-effects emanating from industrial starts (either large-scale, capital-intensive, heavy industrial complexes often utilizing imported technologies or the village and small-scale industries) are concerned. India will continue to[44]:

> "Run the risk of creating petty islands of highly organised western types industries that will remain surrounded by a sea of stagnation. If this fate is to be averted, industrialisation must be so directed, and so complemented by policies in other fields, as to permit simultaneous development outside the sphere of modern large scale industry. "

An argument in favour of industrialisation may run for export promotion of the manufactured goods and products of heavy industry, and importing of foodstuffs. India does possess coal and other metallic ores in substantial deposits needed for a metal trade and could easily penetrate a large share of the world market.

But it will be some time before the industrial products could develop and attain quality standards to compete in the competitive industrial export markets. In other words, premature industrialisation, as a basis for import-substitution, is not a good recipe for rapid economic growth. However, focusing the industrialisation drive upon import-substitution of foodstuffs may have some rationale, "particularly because it minimises the backwash effects of modern industrial growth on established traditional industries," in spite of the fact that[45], "unlike traditional crafts, the agricultural sector is not vulnerable to backwash effects from industrialisation." But the argument has, if any, marginal significance, in an under-developed over-populated economy, with limited material resources plus a subsistence level agricultural sector, "the interrelationship that exists between agricultural productivity and general economic development is such as to stress the urgent need for increased agricultural production," and as I stated earlier, "historically, few countries have managed to industrialise without becoming first self sufficient with respect to agricultural production; hence the oft-cited generalisation that an industrial revolution depends on a prior, or at least a concurrent, agricultural revolution."[46]

Agricultural production and concurrent growth in the industrial sector is indispensable, not only to mechanise the agricultural sector for producing fertilisers, insecticides, and pesticides, but also to release the excess and underutilised labour force now bottled up in the subsistence agricultural sector.

The total number of cultivators and agricultural labourers constitute about 70 per cent of the total labour force, and it has not changed significantly from the 1960s to 1990s. With respect to the reliability of data, there may be some discrepancy if we want to know the under-utilised or disguised employment. But this is an overall outlook of the labour force engaged in the agricultural sector, and the improvement in the labour force depends upon large-scale industrialisation. This transfer of subsistence agricultural labour to industrialisation, concluded El Rashidi[47], may raise:

> "Two major classes of problems with regard to the utilisation of human resources. The first of these is to decide what principles should be followed in drawing up an industrialisation policy that will not only contribute to the economic growth but will also help to raise productive employment. The second is to determine what action will enable new or expanding industries to have the necessary manpower at all levels of skill, in order to ensure that their efficiency and productivity will make for harmonies and continuous development."

We should, therefore, be aware of the fact what Myrdal[48] called a "classical theory that agriculture is subject to diminishing returns whereas increasing returns prevail in industry." In my view, first substantial mechanisation and relatively efficient farming techniques should be employed in agriculture production before shifting the subsistence agricultural labour to industry as illustrated through the flow chart shown earlier in Section (A). This is necessary to cater to mass needs engaged in the industrial sector.

If we consider the Indian economic problem in the light of the existing socio-institutional structure and the attitudinal thought pattern of the conservative society, the high human-resources/land ratio, dimension of the rural population (80 per cent), it seems to me that it is quite rational and logical to emphasise agricultural growth. A realistic agricultural plan has to play a significant role in the general economic development, under the present economic and sociological circumstances. However, the development of an agriculturally related industrial sector to mechanise the largely un-mechanised agricultural sector and to produce fertilisers, insecticides and pesticides is equally important. From this discussion, the following three conclusions can be drawn:

1. Development of agricultural sector with top priority;
2. Development of related industrial sector, primarily to mechanise the agricultural sector and related industries for fertilisers etc.; and
3. Reforms in the socio-institutional structure of the conservative (largely illiterate) society.

Consequently, I have attempted to devise a mechanism (the diagram) to formulate a "balanced" investment strategy based upon "priority". The diagram encompasses all the three variables mentioned above. The first two variables are quite obvious and are shown in the diagram (box 5). The third variable, which I mentioned as the supplementary sector, during the initial stage of growth in both of the sectors, agriculture and industry, should have the following two major objectives:

1. Functional literacy for the farmers for general rural

development, enabling them to understand the modern agricultural techniques.
2. Vocational education within the industry.

Thus analysing the relative contribution of either pursuing the policy of a multitude of sophisticated industrial complexes or increasing the agricultural production and productivity, shows that none of the sectors alone can provide a satisfactory composition of the output (box 6) consistent with the immediate needs and requirements of the people. Both of the sectors have to grow with their relative contribution, and of course, compatible and consistent with the national resources and peoples' requirements. An integrated approach, therefore, for a concurrent development of both the sectors is needed, based upon agricultural priority and related industrial growth, spurred by a supplementary sector (indicated above) under the present economic circumstances of the country: surplus labour, un-mechanised subsistence level agricultural sector, and continuing food shortage.

D. SMALL-SCALE (S-S) OR LARGE-SCALE (L-S) PRODUCTION

"There is an imperative need to evolve a new developmental model in which cottage industries, the small-scale sector and the large-scale industries coexist without encroachment on each other."

—*Prof. Madhu Dandavate*[49], *1989*

The purpose of this section is to review the existing pattern of industrialisation in India with particular reference to small-scale industries as ancillaries to large-scale industries in as decentralised a fashion as possible. The aim is at a balanced socio-economic development to cater to mass needs on an equalitarian basis. As to the policy issue with respect to the small-scale *versus* large-scale industry, I would like to mention the policy of the Congress before Independence under the leadership of Mahatma Gandhi. Jawaharlal Nehru wrote in 1944[50]:

> "The Congress under Gandhi's leadership, had long championed the revival of village industries, especially hand-spinning and hard-wearing. At that time, however, had the Congress been opposed to the big industries, and...it had encouraged this development. ... Any argument as to the relative merits of small-scale and large-scale industry seems strangely irrelevant today when the world, and the dominating facts of the situation that confront it, have decided in favour of the latter."

A comprehensive analysis of the high capital-cost technology, *vis-à-vis* low capital-cost small-scale agro-industries, as the prime mover of the developing agrarian economy of India, is required. The reasons for this analysis are as follows:

1. The food production is falling behind to keep up pace with the compounded interest of human race in geometric progression. This is largely due to low agricultural productivity. Use of fertilisers is one of the main determinants to obtain higher yields per acre, in addition to the quality of seeds, pesticides, irrigational facilities, etc.
2. Agricultural labour in India is hardly employed for more than 6 months a year and the agricultural methods of production have not changed for centuries. Full employment is scarce even in areas where subsistence-farming techniques have been replaced by modern farming and where irrigational facilities exist.

The existing bureaucratic infrastructure for industrialisation for the development of small-scale industries and large-scale enterprises (largely State owned) provides a "balanced" decentralised pattern and thus a strong foundation for growth. The point, however, is that the industrial developmental policies of the government mostly depend upon factors which lie outside the country's existing socio-structural and institutional characteristic features (such as caste-alignments, kinship loyalties, emotional adherence to a tiny patch of land, and thus too traditional to change, etc.), which inhibit the indigenous and authentic development. In other words, development of high capital-cost technologies and large-scale enterprises, without the parallel development of small-scale agro-industries in rural areas (where 81 per cent of the population lives) is simply widening the socio-economic inequality in society. This phase of development is surely against the "spirit" of a socialist pattern of society, which the Government of India claims to establish. The present widening socio-economic inequality in the society indicates that something somewhere is wrong. Either the nature of investment in certain industrial projects is not suitable or the wrong kind of technology must have been adopted. By the latter, I mean that a wrong judgment between labour-intensive and labour saving devices of production must have been made. For the sake of employment, it may be argued to prefer labour-intensive methods of production, despite their being inefficient. Here, I do not necessarily mean that capital-intensive and labour saving production methods must be eschewed altogether. In many instances (such as fertiliser industrialisation, steel production, boiler manufacturing, precision equipments, defence production) high capital-cost technology is the only suitable and feasible method of production. The point is that both the small-scale metal-based and agro-industries must develop abreast with large-scale production methods as ancillaries and in complementarily with, for an accelerated decentralised rural industrial development.

This must be achieved through the development of small-scale industries, associations, preplanned organisation and management of production, a balanced official policy planning, allocation of raw material and other resources to private and public sectors, training of the small-scale industries personnel, provision to provide technological advice to small-scale entrepreneurs which they may not afford due to lack of finance and small nature of their business, guidelines for the marketing facilities of their products, etc. Such a coordinated and integrated pattern can only be achieved through a joint industrial cooperation and an organised collective action among the government manufacturers and semi-government agencies. Unless the political drive is initiated for a balanced and decentralised rural industrial development, and unless rural industrial development through agro-industries form a complementarily with the large-scale enterprises, the ever-widening socio-economic inequality in the poorest among the poor nations can never be overcome.

Now we would like to discuss the techno-economic concept, to develop a kind of technology that maximally utilises the available regional and local resources. A study of comparative advantages and disadvantages between labour saving, labour-intensive and economies of high capital-cost and low capital-cost technologies with regard to specific industry, is required. Then and only then can rural industries be developed and strategies formulated. Having done this, a decision must be made to accept either the so-called "transfer of technology" from the industrially developed nations, or to develop the indigenous methods of design and engineering.

In a country like India, where according to 1971 Census, about 81 per cent of the total population lives in rural areas and about 70 per cent is illiterate for any planning to be realistic, it must take into consideration these two characteristic features of Indian society.

Unless the rural mass energy is released, there can be no authentic development. All the favourable factors and circumstances must be taken into account that might help in developing rural industrialisation. The dominant large-scale industries must emerge through the eventual growth of small-scale agro-industries and rural development. The growth and development of indigenous design and engineering know-how must be the characteristic features in policy planning rather than relying upon the sophisticated technologies borrowed from the industrially developed nations and modifying them to "suit" the needs and requirements of the recipient nation.

Past experience has shown that to rely upon the transfer of technology (from the multinational corporations and their free market economies) as the base to lay down the necessary infrastructure is no more than an illusion, as far as real development is concerned. This transfer of technology from the multinational corporations and its "contribution" to the recipient nations' industrial development becomes in its most passive form when the selfish interests of recipient nations' bourgeoisie, coincide with those foreign collaborations' firms. In K.G. Charles[51] words,

> "When there is an unholy alliance between the domestic ruling feudal class and a handful of powerful foreign business interests, these groups can exercise a stronghold on development. The motivation then exists to evolve improved technologies."

While discussing the production of fertilisers, it seemed that a complementary development of small-scale and large-scale enterprises was needed to cater to the rural masses' needs. Regarding the so-called "appropriate" or the "intermediate" technology, Charles[52] said that it is more effective and viable than the traditional technology it will replace. Its aim must be to fulfil the needs of its masses.

Therefore the developmental policy must be so balanced as to fulfil the routine operational equipment requirements of large industries and developing the agro-industries and metal-based small-scale industries. The policy planners must also take into consideration the social and institutional structures in a traditional society that are inhospitable to innovations. In any case, the coordination between small-scale and agro-industries is of paramount economic interest for a decentralised development of economy and an integrated rural reconstruction.

The strategy for accelerated industrial development for small-scale industries in close complementarily with large-scale industries must provide the necessary infrastructure for rural agro-industrialisation. The development of such small-scale industries need not necessarily be concentrated in urban areas. This will then reduce the concentration and formation of slums in the large metropolitan areas. The strategy for a complementary development of small-scale and large-scale industries neither underestimates nor over emphasises the 'contribution' of large-scale industries to the overall national developmental policies. This complementary development is mutually advantageous for a variety of reasons. In India, the most neglected sector of economy is rural agro-industrialisation. Undoubtedly, this is where its development is immediately needed to generate employment for the rural illiterate and unskilled masses. Agro-industrial methods have always been on subsistence levels and have not changed for centuries. Exceptions to this are only the training facilities in research institutions. Most under-employment and unemployment is in rural areas where about 80 per cent of the population lives. Rural agro-industrialisation in India is required for certain other reasons. Agriculture in India does not provide jobs for more than half of the year. Even in areas, where subsistance-farming techniques have been replaced by modern farming and where adequate irrigational facilities exist, full employment is scarce.

Certain small-scale industries in rural areas, not requiring highly intensified technologies (such as agricultural implements, oil processing, cotton ginning, fruit and vegetable processing) can be easily established. The development of such agro-industries in rural areas will bring a revolutionary change, without much radical reforms.

These agro-industries in rural areas will enable the banking and other commercial agencies to have their offices there. It will mobilise the rural masses and thus the abundant mass energy may be release. As a result, the other infrastructure for rural development will follow (such as roads, railways and the means of communications). Its by-product will be a slow down in migration from rural areas. Agro-industries in rural areas will provide an incentive for skilled and educated unemployed workers in urban areas to seek employment in rural areas. With this kind of decentralised industrial development, the rural masses will no longer be requiring the "services" of the middleman and that of intermediaries—the exploitative sector of economy.

The launching of massive agro-industries in the rural areas will then transform the existing lop-aided agricultural economy into a more balanced agro-industrial economy.

To initiate methods and techniques of rural industrialisation certain pilot projects of promotional nature may be launched, including the provision for training facilities, common service facilities, technical assistance, marketing techniques and credit for the purchase of machinery and construction for work sheds, buildings, etc. The pilot projects for agro-industrial development may also include animal husbandry products as well as industries based upon forests such as pulp and paper, saw mill, fibre raw material wood-based panel industries, and so on. These types of small-scale industries are usually labour-intensive and relatively use low capital-cost technologies and cheaper equipment. The establishment of these small-scale industries as ancillaries in rural areas will provide a balanced economic growth between rural and urban areas in as decentralised a fashion as possible. It will generate employment on a massive scale and provide at least a part of the solution to the ever-increasing problem of socio-economic instability.

Here, it would be pertinent to make a comparative analysis of the economies of high capital-cost and low capital-cost technologies. India is a country short of capital, with low-skills, but abundant human resources. Therefore, for any planning to be successful and any technology to be realistic, it is necessary that it must release the mass energy. The substitution of labour for capital in manufacturing technologies is imperative to create widespread job opportunities. Imported technology cannot be viewed as a panacea to rapid industrialisation and widespread exploitation of natural resources, without being modified to match the recipient nation's own infrastructure and value systems. The adoption of this imported technology has created sophisticated industrial complexes on one hand, and, on the other, there is a widespread under-employment and unemployment particularly in the rural areas. In addition to this, there is a loss to the overall national economy, when capital-intensive manufactured goods compete with indigenous goods (this is so because high-cost technology produces relatively more fine quality of goods than low capital cost technology). Also high capital-cost and sophisticated technologies are usually labour saving, and so create relatively few jobs. Many of these jobs require highly skilled labour, which inhibit the prospects of rural people for employment.

In case of the fertiliser industry, however, as stated earlier, the high capital-cost technology and sophisticated manufacturing processes are more effective and suitable. They are also economical in operation. Therefore, in this case, capital-intensive production simply cannot be eschewed. So also is the case with other industries, such as production of steel, boilers, precision equipment, defence material, nuclear power generation equipment etc. Thus it seems that, for a balanced and decentralised industrial modes of production, a collective action for the development of small-scale industries as ancillaries to large-scale industries in a coordinated and integrated way, would be more feasible and suitable. It has two major advantages. First, lack of institutional infrastructure suffered by most of the small-scale industries can be minimised and, secondly, the governmental bureaucratic institutions can cope effectively with the numerous small-scale industries. Small-scale industries in this way can help provide an accelerated industrial development along with medium and large-scale industries. Collective action may be between manufacturers and government and semi-governmental agencies. From such an action, joint services and common facilities for marketing, purchasing, production advertising, research and development, advisory services, training for industrial personnel, export promotion, and other types of inter-firm links which exist in the ancillary and the sub-contracting industries, can be more effectively and efficiently controlled. However, one of the major requirements for such a collective action to be successful is that the governmental institutions (such as productivity centres, industrial research centres, small industries institutes and similar organisations) must create favourable conditions to maintain the environment. They must organise the small-scale industries to help resolve their problems rather than merely give a theoretical "advice" to maximise the profits. With this development of industries through cooperatives, a decentralised industrial development will eventually emerge.

Looking at the present situation, the picture may not be as discouraging as it would appear, if lesson are learnt from some of the past failures and concerted efforts are made to analyse their causes. India has a very high degree of institutional infrastructure for industrialisation. The need is for an effective governmental control, coordination, and integration between requirements and capabilities of the products of the small-scale industries and large-scale industries.

In any case, with mutual industrial cooperation and organised collective action, the role of the government must be secondary. The collectives and cooperatives of the industries must themselves determine what they can do within the limits of the human and material resources at their disposal. The role of the government must be only to supplement their efforts through financial aid and other technical advice, which the small-scale industries may not afford individually.

The complementarily of small-scale and large-scale industries can only be achieved by the development of small-scale industries through cooperatives. The cooperative organisations can play a substantial role in promoting the growth of small-scale industries. Cooperative action must be an "ally" in any sector of economic activities, especially in a developing economy. Cooperatives in India mainly consist of "service cooperatives" which arrange for the production of semi-processed raw materials for entrepreneurs who run small-scale industries of their own. There is a National Federation of Industrial Cooperative Societies, which provides guidance and coordinates the various managerial activities. The establishment of industrial cooperatives is particularly useful to agricultural producers in securing better returns for their labour and their products. These cooperative societies assist in increasing the efficiency of the workers, procure raw materials on wholesale basis, and supply to individual craftsman, and so on.

Thus to formulate a strategy for the complementarily of the small-scale industries to the large-scale industries in a decentralised fashion to cater to mass needs, particularly the rural unskilled and illiterates, a multivariate interaction between the factors of production (such as property, available machinery, tools, capital, etc.) and the available technical means of production (the indigenous technical know-how and modification of the imported technology) is a virtual necessity. For balanced decentralised economic development, both the small-scale and the large-scale industries must develop on an interdependent pattern of growth.

The Industrial Policy Resolution of 1948 stated the importance of small-scale industries, particularly to achieve local self-sufficiency and better utilisation of local resources. The object was to provide the basic consumer goods. During the First Five-Year Plan, the Government of India and the Ford Foundation jointly sponsored the visit of the International Perspective Planning Team in 1953-54. As a result, four regional institutes in Calcutta, Bombay, Delhi, and Madras were set up to provide technical assistance and guidance for small-scale industries. For assisting the marketing of products, a National Small Industries Corporation was also established. The total outlay, however, for the development of small-scale industries was only Rs.50 million in a Plan of Rs.42, 000 million, less than 0.1 per cent. This percentage and amount of outlay to develop small-scale industries indicate the low priority the planners gave to the growth and development of small-scale industries.

The Second Industrial Policy Resolution was enunciated in 1956. The official policy planning is fundamentally based upon the socialist pattern of society and nationalisation of heavy industries requiring high capital-cost technologies. The respective roles of the public and the private sectors of the industry were outlined in a statement on Industrial Development Policy read in Parliament on April 30, 1956.

Now, since the mid-1980s, the policies have begun to change to liberalise the economy. This liberalisation has further been accelerated from 1991-92 under the prime minister-ship of P.V. Narasimha Rao and his finance minister Dr. Man Mohan Singh.

This accelerated liberalisation of the economy is more likely to benefit the industries in the large- and medium-scale sectors. The interests of the entrepreneurs in the small scale industrial production are bound to suffer for the generally capital starved low-technology based small scale industrial sector in India can not compete with the medium- and large-scale industrial sector in open free market. So some sort of stringent measures by the State to protect the interest of the small-scale sector are required against the otherwise inevitable onslaught by the medium- and large-scale sectors on the small-scale sector.

During the Second Five Year Plan, the outlay for small-scale industries increased to Rs.610 million out of a total Plan of Rs. 60,000 million, just about 1 per cent (an increase more than ten times over the amount assigned in the First Five Year Plan). In July 1967, an Industrial Licensing Policy Inquiry Committee was appointed to "prevent" the growth of capitalistic modes of production in the private sector. As a result a new licensing system was announced in 1970.

Undoubtedly, there has been a remarkable progress in range and sophistication of products manufactured by the small-scale industries. But this has largely been in metal-based small-scale industries, where its production counts as much as 60 per cent of the total production. As far as the development of the small-scale agro-industries for rural industrialisation is concerned, the progress is just the opposite.

Any attempt to surmount the inherent difficulties in promoting the role of the small-scale agro-industries as ancillary units to large-scale enterprises, and its potential contribution to an agrarian under-developed economy like India, is justifiable. In addition to this, the abundant rural population of India poses certain peculiar social-structural and institutional obstacles to achieve the economic goals. The performance and implications of industrial policies must not be allowed to depend on factors outside the social-structural and institutional organisations of the country; otherwise there can be no decentralised and authentic rural agro-industrial development. Such variables as income distribution and class-analysis (caste alignments and kinship loyalties of a traditional society) must always be considered in policy planning system. To employ even indigenous methods, in spite of their being inefficient, may be argued for the sake of reducing unemployment.

Local indigenous methods of production must be closely linked and integrated with the scientific organisations and the science and technology policy-making system. Then and only then transfer of technology from the developed nations can help India to break out of the straight jacket and in a real sense be accelerating and developing the indigenous technological methods of the recipient nation. To "modify" the borrowed technology to suit the needs of the recipient country, fitting into her own factors of production, without considering the indigenous modes of production, is totally absurd, irrelevant, and wasteful in many instances.

The ever-widening socio-economic inequality in the society indicates that something somewhere is wrong in terms of industrial development. It may be either in the nature of investment or a wrong kind of technology must have been adopted in developing the industrial infrastructure.

In the light of above analysis, it seems that a functional relationship be established between technology and employment to mobilise the hitherto stagnant rural masses of India. In other words, the adaptation of any kind of technology should not produce certain theoretical approaches alone. The two highly interchangeable aspects (theory and methodology) are one and the same thing; "Appropriate" technology must have a definite relationship and a functional validity. At times, the appropriate technologies may be quite inappropriate only because of the lack of or weak communications. The choice of technology can be inappropriate in a number of ways as summarised by Stewart, (1) through wrong selection system, (2) through weak communications system, (3) through expensive technology transfer, (4) through inadequate work knowledge. Only the "appropriate" methods themselves are not appropriate unless the inputs, particularly of machinery essential for adopting them, are available.

Small-scale industries, while having complementarity with

large-scale enterprises, are also the main sources of stimulating indigenous entrepreneurship. Individuals from different sectors of society, particularly the rural and unskilled, can only learn and realise their drive towards entrepreneurship through employment opportunities in small-scale industries. This is particularly true because (1) there are not enough large-scale enterprises in India, and (2) whatever there are, they seek only high trained and skilled people for their management positions.

Small-scale industries also facilitate the process of import-substitution because large-scale industries usually require equipments, components and parts, to replace those are and worn reasons. Small businesses can manufacture these parts more economically.

The Government of India has to assume a special "responsibility" for the development of small-scale industries and provide a sophisticated institutional infrastructure for this purpose. This include establishment of a number of financing institutions for credits, reservation of certain quotas of raw materials to small-scale industries, training of industrial personnel, prototype training centres, and the formation of small-scale industries associations.

This role of the State to support and assist the small-scale industries is highly desirable.

The smallness of enterprises and limitations of the individuals who own them (because of lack of adequate finances and low entrepreneurship skills) are major constraints in the independent growth of small enterprises. The role of state is seriously lacking in working out a policy and programmed for industries in both small-scale and large-scale industries. This has resulted in their slow growth.

The importance of the growth and development of small-scale industries has finally been recognised. After the general election in 1989, the new government in India was formed under the prime ministership of V.P. Singh on December 2, 1989.

Prof. Madhu Dandavate[53], the then Finance Minister of India, soon after taking office in 1989, stated:

"Cool and sedate statisticians complacently quote figures of the growth rate of our economy to assess the degree of development. However, the question we must ask ourselves is "growth—yes; but growth for whom?" In a country where vast sections live below the poverty line, the developmental process must not be like snowfall wherein snow is accumulated on high mountain peaks, but like rainfall wherein water rushes to the grassroots level. ...Gains for the affluent through favoured treatment, and gains for the poor through mercy cannot be the guiding philosophy for an India yearning for distributive justice.

There is an imperative need to evolve a new developmental model in which cottage industries, the small-scale sector, and the large-scale industries co-exist without encroachment on each other; so that those decentralised industries particularly in the rural sector, providing employment to the rural masses, can be augmented. ... Under this new model of development, commodities which can be produced in cottage industries will not be encouraged in the small- scale sector, and those which can be conveniently manufactured in the small-scale sector will not be encouraged in the large-scale sector. This, of course, will have to be implemented in a phased way, without disturbing existing structure and capacity."

Since 1991, when the so called economic reforms began with full swing, the net outcome of this more than a decade of reforms from 1991 to 2001 is well summed up by Mohan Guruswami, the then adviser to the Finance Minister of India:[54]

- Industrial growth in 1981-91 was 7.7 per cent. Since 1991-92 it has been 5.8 per cent.
- Agricultural growth, which averaged 4.04 per cent annually for the decade 1981-91, fell to 2.3 per cent annually for the period 1991-92.
- The annual electricity generation capacity growth fell to 3.5 per cent during 1991-99 as opposed to 12.4 per cent for 1981-91. Electricity generation growth fell

from 17 per cent to 7.8 per cent for the corresponding period.

- Trade deficits have been growing exponentially. In 1980-81 the adverse balance was Rs. 5,838 cr. In 1990-91 this was Rs. 10,645 crores. In 1998-99 it swelled to Rs. 55,478 crores.
- Poverty levels have remained constant at about 33 per cent. Therefore, the numbers living below the poverty line have swelled by 46.7 million to 326.04 million in 1999.
- Rupee: in 1981 the Dollar was worth Rs. 8, in 1991 this changed to Rs. 18, and now to almost Rs. 44.

NOTES

1. Mahatma Gandhi, *Collected Works*, Vol. 64, (New Delhi; Ministry of I & B, 1976), p. 73.
2. *Ibid., Collected Works*, Vol. 71, pp. 424-25.
3. *Illustrated Weekly of India*, December 31, 1989; p. 15.
4. By the term "Mass Production" what I mean is that the consumer goods should be produced, and then be available, to all the people in the society. By the term 'production-out-of-masses" what I mean is that the consumer goods should be manufactured as far as possible by utilising the abundant human resources.
5. M.I. Volkov (Ed.), *A Dictionary of Political Economy* (Moscow: Progress Pub., 1985); p. 14.
6. Mahatma Gandhi, *Collected Works* (New Delhi: Ministry of I&B): Volume 74, (1978), p. 169; and Volume 66 (1976), p. 266.
7. *Ibid.*, Vol. 64 (1976), p. 73.
8. V.I. Lenin, *Alliance of the Working Class and Peasantry* (Moscow; Progress Pub., 1965), p. 123.
9. Karl Marx, *Das Capital*, Vol. I (Moscow: Progress Pub., 1954), p. 337.
10. Jawaharlal Nehru, *Discovery of India* (London: Marion Book, 1946), pp. 535-36.
11. Mahatma Gandhi, *op. cit.*, Vol., 71 (1978), p. 56.
12. Mahatma Gandhi, *Collected Works*, Vol. 69 (New Delhi: Ministry of I&B, 1979), p. 53
13. *Ibid.*, p. 53. This concept of "trustee" is based on my interpretation on Gandhi's ideas as noted in pp. 53-69.
14. Doreen Warriner, "Land reform and economic development" in

Eicher and Witt, (Eds.), *Agriculture in Economic Development* (New York; McGraw Hill, 1964), p. 281.
15. Karl A. Wittfogel, "Communist and Non-communist Agrarion Policies" in W. Danglas Jackson (Ed.), (Seattle; University of Washington Press, 1971) p. 65.
16. *Ibid.*, p. 65, Comments by Werner Klatt on the above paper.
17. Even J. Long, "Economic Basis of Land Reform". In Tara Shukla (Ed.), *Economics of Agricultural Development* (Bombay: Vora Pub., 1969), p. 291.
18. See G. Tones, *The Role of Science and Technology in Developing Countries* (New York: Oxford Uni. Press 1971); p. 70.
19. Lester B. Pearson, *Partners in Development* (New York: Praeger Pub., 1969), p. 24.
20. G. Myrdal, *The Challenge of World Poverty*, (New York: Pantheon Books, 1970); see Chapter 4.
21. G. Jones, *op. cit.*, p. 43.
22. Quoted in *Ibid.*, p. 69.
23. Discussed later in this chapter.
24. *Ibid.*
25. Investment in R & D (agriculture) has been already put in box (8) to get higher yields per acerage. The output of this, when fed back to basic industries (box 7) achieves initial stages of heavy industry (box 10). After this, we spend money on R & D (industrial) (box 11).
26. David E. Lindstrom "Developing from and Farm Policy in Sweden" in C. Zimmerman and R. Du Wor (Eds.); *Sociology of the Underdeveloped* (Calgary; The University of Calgary, 1970); p. 356.
27. For detailed analysis, *see* Sures Chandra Jain, *Education for Socio-Economic Development*, Concept Publishing Co., New Delhi, 2004.
28. W.Y. Yang, *Methods of Farm Management Investigations* [Rome; Food and Agricultural Organisation (FAO) of the United Nations (U.N.) 1965]; p. 124.
29. L. Stephnov, "Why social reforms are in evitable" in Y. Zhukov, *et al.*, (Ed.), *The Third World* (Moscow: Progress Pub., 1970); p. 137.
30. For detailed analysis, *see* Sures Chandra Jain, *Globalization, Debt, Arms and Reforms in South Asia*, Concept Publishing Co., New Delhi, 2004.
31. L. Stepnov, *op. cit.*, p. 134.
32. Paul R. Ehnlich, *The Population Bomb* (New York: The Ballantine Books, 1971); Cover page citation.
33. *The Hindustan Times*, New Delhi; 23.8.1989; p. 11.
34. Barrington Moore, Jr. *Social Origins of Democracy and Dictatorship* (Boston; Beacon Press, 1966); p. 407.
35. G. Myrdal, Asian Drama; *An Inquiry into the Poverty of Nations*, Vol. II (New York; Pantheon Books, 1968); p. 1151.
36. For details of this issue, see W. Nicholls, "Agricultural Policy; the

place of agriculture in economics, in Kenneth Berrill, *Economic Development* (London; McMillan, 1968); p. 359.
37. *Ibid.*, p. 366.
38. *Ibid.*, p. 367.
39. *Ibid.*, p. 367.
40. G. Myrdal, *op. cit.*, p. 1150.
41. Rene Dumont, "A World Gone Mad", in CERES FAO of the U.N., Rome, Nov.-Dec. 1973; p. 17.
42. W. Malernbaum, "India and China; Contrasts in Development" in *American Economic Review*, 1959, Vol. 49; p. 300.
43. W. Nicholls, *op. cit.*, p. 363.
44. *Ibid.*, pp. 361-62.
45. Myrdal, *op. cit.*, pp. 1204, 1244.
46. *Ibid.*, p. 1249.
47. Fathi El. Rashidi, *Human Aspect of Development* (Brussels, Belgium: I.A.S., 1971); p. 146.
48. G. Myrdal, *op. cit.*, p. 1156.
49. Professor Madhu Dandavate became Finance Minister in December 1989 in the new government under the prime minister ship of Mr. V.P. Singh. The comments of Professor Dandavate appeared in the *Illustrated Weekly of India*, December 31, 1989; p. 15.
50. Jawaharlal Nehru, *Discovery of India*, (London: Marion Books, 1946); p. 409.
51. K.J. Charles, "Intermediate Technology" in *Cooperation Canada*, Canadian International Development Agency (C.I.D.A.) Ottawa, July–August, 1974; p. 7.
52. See *Ibid.*
53. *The Illustrated Weekly of India*, Bombay, December 31, 1989; p. 15 Footnotes.
54. Mohan Guruswamy, The sad face of liberalisation: After almost 10 years, all key indicators tell us that the New Economic Policy of 1991 has decelerated India's growth, *Indian Express*, April 11, 2000, p. 11 (Former advisor to the FM).

3

Land Reforms

"The countries that have succeeded in modernizing their economies all began with far-reaching land reform. In our country, the major obstacle in the way of meaningful land reform has been the lack of firm political will at the highest levels."

Editorial, The Hindustan Times[1], *1992*

...Put simply, people are poor for want of entitlements, physical and/or human. Lack of access to agricultural land, paucity of work on land (wage work), and non-availability of education (for enabling occupational diversification) or shortage of capital is the fundamental causes of poverty. Thus, in regions where agrarian reforms have been less than successful, agriculture has been relatively stagnant, infrastructure development is less or social/educational deprivation (including that based on caste) is high, poverty is high, and vice versa.

National Institute of Rural Development[2], *1999*

A. LAND REFORMS: A STRUCTURAL PROBLEM IN THE FORMATION OF AGRO-INDUSTRIAL COMPLEX

It is evident from the three preceding chapters (Chapter 2, agricultural mechanisation and rural industrialisation, Chapter 3, utilisation of natural resources, and Chapter 4, mobilisation of human resources) that the development of agriculture is the basis for rural development, and that it would serve as an infrastructure base if the industrialisation of the nation were to occur. The development of agriculture is a fundamental requisite not only for the process of rural development, but also if we are to remain self-reliant, depending on our own material resources, and avoiding the dependence on external aids, loans, etc. National Institute of Rural Development, in *its India Rural Development Report, 1999*, mentioned about the impact of land reforms that:[3]

> Land Reforms' have been viewed as the core component of the poverty alleviation strategy in India. Most of the analytical studies on rural poverty could not, however, establish the significant influence of 'Land Reforms'.
>
> The apparently non-significant effect of 'Land Reforms' on poverty could be ascribed to the pace and quality of these institutional reforms in several states. The benefits of reforms may not be adequate to enable the poor to cross the poverty line due to the meagre size of land, besides its (inferior) quality. Where these have received state support, the poverty levels declined considerably.
>
> In 20 out of the 58 NSS regions where poverty levels were low in 1993-94, the land holding structure was favourable in 13 regions suggesting that if land reforms could influence the landholding structure, then poverty could be contained. Thus, more emphasis needs to be laid on vigorous enforcement of these institutional reforms and also on the quality of implementation process. Land based activities under the IRDP should receive extension-cum-credit support to make a sustainable dent on poverty reduction.
>
> ...Put simply, people are poor for want of entitlements, physical and/or human. Lack of access to agricultural land, paucity of work on land (wage work), and non-availability of education (for enabling occupational diversification) or shortage of capital is the fundamental causes of poverty. Thus, in regions where agrarian reforms have been less than successful, agriculture has been relatively stagnant, infrastructure development is less or social/educational deprivation (including that based on caste) is high, poverty is high, and *vice versa*.

Indeed, the development of agriculture would depend on the restructuring of the existing land system. Regarding the restructuring of the land system, we illustrate the way it was in ancient times in India, and its importance in contemporary India as outlined by Mahatma Gandhi and even by the Jawaharlal Nehru. That Nehru did nothing to institute the re-structuring of the land-system, is a different matter. (This would be quite evident later in this Chapter.)

According to Karl Marx[4], (*Das Capital*, Vol. I, 1868):

> "Those small and extremely ancient communities are based on possession in common of the land, on the blending of agriculture and handicrafts. The land is tilled in common, and the produce divided among the members."

The following is the way Mahatma Gandhi[5] advocated the issue of the land system as early as 1940:

> "In reality the toiler is the owner of what he produces (1936) ...who can dispute the fact that the grinding poverty of the masses is due to their having no land that they can call their own ?"

Jawaharlal Nehru[6], in his *Discovery of India*, written in 1944, said:

> "A democratic collectivism need mean an abolition of private property, but it will mean the public ownership of the basic and major industries. It will mean the cooperative and collective control of the land."

We now outline the policies of the government in the post-Independence era regarding land reforms, dealing with objectives and programmed of action, abolition of intermediary tenures, tenancy reform measures, land ceiling measures, and consolidation of holdings[7]:

> "Immediately after Independence, land reform measures were initiated in the country. The objectives of the land

reforms policy were set out by the planners as the removal of such institutional and motivational impediments to the modernisation of agriculture as were innate in the agrarian structure inherited from the past and the reduction of gross inequalities in the agrarian economy and rural society which stemmed from rights in land.

These objectives were translated into the following programmes of action: (a) the abolition of the prevalent intermediary system between the state and the tiller of the soil; (b) conferment of ownership rights on the cultivating tenants in the land held under their possession; (c) imposition of ceiling on agricultural land holdings as a measure contributing to the modernisation of agriculture and to eliminate parasitic absentee landlordism; (d) rationalisation of the record of rights in land so as to make it reflect the rights of tenants, sharecroppers, and other categories of insecure landholders; and (e) consolidation of holdings with a view to making easier the application of modern techniques of agriculture. Provision of dependable irrigation facilities, etc., was put down as one of the main items in the programme of action."

The objectives and programmes of action concerning land reforms, as outlined in the passage are noble, indeed, but how far these programmes of action are actually implemented ?

The National Sample Survey of 1961-62 provided the data regarding the estimated number of operational holdings and geographical area operated by the size of holding, as shown in Table 3.1. An operational holding covers all kinds of land whether entirely or partially for agricultural production purposes. The corresponding 1970-71 data regarding the estimated number of operational holding and geographical area operated by the size of holding are shown in Table 3.2.

From Table 3.1, we notice that by 1961

(a) 19 per cent of holding were small (1 to 2 hectares), covering 12 per cent of the area.

(b) 15 per cent of holding were semi-medium (2 to 4 hectares), covering 19.5 per cent of the area.

(c) The rest, *i.e.* 66 per cent holding were medium/large (4 to 10 hectares), covering 60.5 per cent of the area.

Table 3.1: Estimated Number of Operational Holdings and Area Operated by Size of Holdings, 1961-62

Area of Holdings	*Number*		*Area Operated*	
	Million	*Per cent*	*Million Hectare*	*Per cent*
Less than 1 hectare	19.8	39.1	9.2	6.9
1 to 3 hectare	18.0	35.5	32.1	24.1
3 to 5 hectare	6.1	12.0	23.0	17.2
5 to 10 hectare	4.5	8.9	30.6	22.9
10 to 12 hectare	1.8	3.5	23.1	17.3
12 hectare and above	0.5	1.0	15.5	11.6
Total	50.7	100.0	133.5	100.0

Source: Charan Singh, *Economic Nightmare of India* (Delhi: National, 1981); p. 127 National Sample Survey, 17th Round. An operational holding cover all kinds of land used wholly or partly for agricultural production.

We have illustrated the situation in 1970-71 as evident from Table 3.2. The number of operational holding in 1961 was 50.7; in 1971 it rose to 70.5. In 1961, there were 39 per cent of holding under one hectare each; in 1971 the number rose to 51. In 1961, the number of farms with 10 per cent or more hectares each was 2.3 million; in 1971 it rose to 2.8 million, the average area being 17 to 18 hectare. The number of large forms in 1961 was 38.6 million, covering approximately 28.9 per cent of the land; in 1971 it rose to 50 million, covering approximately 37.5 per cent of the land.

There have been land reforms all through the decade 1961 to 1971, but these reforms have almost always benefited the uppermost rung of peasantry over the lowest rung. This is not the state of affairs that is being practiced in India.

The same trend, which we have just outlined during the decade, from 1961 to 1971, continued during the decade, 1971-1981. The same is also true during the 1990s[8].

"In our country, therefore, (a) where it is land, it is the limiting factor, not labour; (b) where the area of land a cultivating family (usually consisting of two workers) holds on an average, today amounts to bare 6.00 acres or so; (c) where the rate of population growth is very high or development of non-agriculture is proceeding at such a slow pace that the man: land ratio of the farming population is going down instead of going up, it is in the interest of the people that:

(a) A ceiling on present possessions of land is imposed at a level not more than 27.5 acres per adult worker (including, of course, his wife and minor children, if any) and the area that consequently becomes surpluses distributed to those who possess no land at all or possess less than 2.5 acres each;
(b) A floor is laid at 2.5 acres, that is, if possible, the law relating to transfer and partition of land in future is so amended that the area of land per worker is not reduced below 2.5 acres; and
(c) Future acquisitions of land are so regulated that, along with what he may be already possessing, the total area a man comes to hold does not exceed a particular limit (which may be fixed somewhere between the ceiling and the floor)."

Now in the light of the statement by Jawaharlal Nehru[9], that in India what we need is the cooperative and collective control of the land and the policies on land reforms we have just outlined, we go on to analyse what Nehru actually did in this regards during his virtually unopposed rule from the moment of Independence in 1947 to the time of his death in 1964. We will analyse what is obviously the utter failures of Nehru to implement what he proposed that got him into power, with a case study of the province of Bihar, when a movement in that province was going on in 1974 under the leadership of Jaiprakash Narain[10], a widely revered leader of the nation. The case of Bihar, for all practical purposes, is a representation of the national malaise. The motto of the nation's Constitution is a 'socialist' form of society; in essence, it is out right opportunism of the rulers, or an outright capitalist agrarian structure[11] in the name of socialist pattern. Let us now have a look at the structure of peasantry in India. Frantz Fanon[12] said that:

> "The peasant who goes on scratching out a living from the soil, and the unemployed man who never finds employment, do not manage, in spite of public holidays and flag, new and brightly coloured though they may be, to convince themselves that anything has really changed in their lives."

The division of peasantry for the purpose of class-analysis can be made on the basis of structural differentiation, defined by the relations of the means of production to the other monetary resources they control. Thus a peasant society can be categorised into three main groups:

1. Landless peasants,
2. Large landowners, and
3. Middle class peasants.

Now such a classification of peasantry, based upon the difference in their wealth and means of production on a linear scale, requires that these modes of production be defined. In other words, essential distinctions are those of the modes of production, and not those of relative differentiation in wealth or property. A definite relation, therefore, can be made between particular classes of peasant and their mode of production as follows:

Landless Peasant

This class usually comprises paid wage earners. It includes farm labourers, agricultural proletariat or the sharecroppers. It is exploited class of the peasantry. They work for wages or crop shares in exchange for labour, usually for the large landowners and capitalist farmers. Landless peasantry, in other words, is at the bottom of the heap, and therefore is a potential source of revolt and insurrection against undue exploitation by the large landowners and capitalist farmers. Ferment develops when their economic conditions reach the ironic state of 'sub-human life and superhuman labour'. The starving peasant outside the class-system is the first among the exploited to discover that only violence pays[13]. If revolutionary ferment is unorganised, its consequences might be serious, leading to further oppression of the peasantry. If organised and successful, it could change their life, as it happened in Tsarist Russia in 1917 and in Imperial China in 1949.

Large Landowners

This class of peasantry usually owns a substantial amount of land. Rich farmers or capitalist farmers belong to this class. The basic difference between landowners and capitalist farmers is that the latter may, or may not; participate in the cultivation process, while the farmers usually do. They either employ wage-labour (agricultural proletariat) or sharecroppers (the landless peasantry) described above. Their farming, therefore, is largely based upon the exploitation of the landless peasantry. This class is economically as independent on landless peasantry as the peasantry is dependent on them (landowners) for their livelihood.

Middle Class Peasants

This is a "critical" class of peasantry. It stands in between the two mentioned above. This class, for theoretical purposes, owns sufficient land for their requirements. They do not exploit the landless peasantry, nor do large landowners exploit them. However, certain deviations do occur and the seasonal labourers employed by them are but a "transitional " period, said Lenin.[14]

Class-Analysis

Now let us look briefly at Marxist-Leninist ideology concerning peasantry and workers and the relationship between the two.

The division between workers and peasants is a result of the rise and growth of the Marxist Socialism. But must the peasants have leadership from outside?

In the Eighteenth Brumaire, Marx explained that the peasants formed a class in so far as, in their economic organisation and in their life, they differed from both the bourgeoisie and the proletariat, and were antagonistic to them.[15] Politically they lacked the unity of a class, being rather an agglomeration of individuals, which he compared to a sack of potatoes. Again, why did they choose the working class as the ruling class or the dictatorship of the proletariat? Lenin emphasises that these are the only two classes, which do not exploit. On May 26, 1921[16], he said that for the "first time in modern history we have a social system from which the exploiting class has been eliminated but in which these are the only two different classes."

However, while advocating the dictatorship of the proletariat, Lenin emphasised that there should be a definite relationship between the working class and the peasantry. Lenin emphasised that the organisation of the population must be through cooperative societies. The transition from the old system to the new one should be the simplest, easiest, and most acceptable to the peasant. Workers must help the peasants in their fight for land and also must help them to formulate the land question in straightforward, clear, and definite terms. The peasants, like the petty bourgeoisie in general, occupy an intermediate position even under the dictatorship of the proletariat. Therefore, Lenin advocated that, even if the decrees of the government are right, it is wrong to impose them on the peasants by force. The special form of alliance between peasants and workers, which Stalin had analysed in "Problems of Leninism", and which had been embodied in the Final Status of the Comitash (1928), was based on the crucial point that proletariat must never surrender its revolutionary leadership, equally on the view that collectivisation would have to be a gradual process. Engel's had assumed that it would take generations to achieve this, and Lenin had insisted that, in the eventual progress from individual to specialised agriculture, the small peasant must be persuaded, not coerced[17]. Stalin maintained that, while the workers cannot share power with the peasants or any other class, the peasants need not be denied a voice in directing their own fate.[18]

In my view, while making this class analysis between peasantry and workers, and since peasantry lacks the unity of a political class because it is an agglomeration of individuals, the notion of the dictatorship of the proletariat, regarding the emerging nations in the aftermath of colonialism, simply means that these nations must skip the stage of capitalism and pass straight on to socialism. This is necessary, if these new nations[19] have to avoid the class differentiation. As we see in India, with a case study of Bihar, the virtues of the socialist pattern of society were being preached right after (an even before) Independence, but in fact they have never really been practiced. As a result, it is obvious there is an ever-widening economic disparity between rich and poor in one of the poorest countries on earth. The appalling misery of a poor Indian is beyond imagination. In the CBC National News at 2300 hours, in Ottawa, on October 30, 1974, I unfortunately watched the poor peasants in thousands, who were simply left to die of starvation. They were too fragile to get to government and relief agencies—operated kitchens. Can this happen in a socialist society?

I must agree, however, that skipping of this capitalistic stage in the aftermath of colonialism is not an easy task, at least in the initial stages of development. This is, because it is a system, which the colonialists operated. It has its roots deep in the national political economy and is in the very core of its decision-making system. Nonetheless, a constant and relentless struggle, slow and gradual though it may be, will follow the road to socialism. But it is not possible, unless the nation's-bourgeois and ruling upper class want to give up their own self-interest. They simply have worn white masks over their native skins, gifted in most cases by the outgoing colonialist.

While wars change the external character of a nation, it is revolution that brings about the necessary internal reforms and changes. The latter is a phenomenon that is required to cope with the existing ferment in the society arising out of a number of socio-economic factors, such as unequal distribution of wealth, unequal opportunities for its citizens for growth and advancement, discrepancy between means of production and items of consumption, corruption at its staggering heights, and other such socio-economic factors. The idea of revolt, civil war, insurrection, or revolution becomes even more acute and pressing when the existing socio-economic structures begin to function in the name of self-seeking bureaucracy in the interest of the ruling class coalition and its alleged association with the national bourgeoisie and large landowners. This is exactly what is happening today in virtually the whole of the Indian sub-continent. The anger of the masses only culminates in revolt and revolution, when the ruling class, the elite, control the means of production and consume the subsequent items of consumption so that there is virtually nothing left to the working class and the peasantry worth a human living. In other words, ruthless exploitation is generally essential before the latent anger can be transformed into a full-scale revolution.

Putting up hindrances in the way and blocking the alliance between landless peasantry and the exploited working class, is an outright betrayal of implementing the promised agrarian reforms for a socialistic pattern of society. A revolutionary ferment can only be delayed, but not averted.

Stalin concluded that[20], "the peasant question of the ally of the proletariat in its struggle for power, is a derivative question". The "peasant" question is one of the vital problems of Leninism[21]. The twelve years of constant struggle between the two revolutions in Russia, in 1905 and the final victory in October 1917, is a history of constant struggle between the bourgeoisies the ruling elite and the proletariat and peasantry. After the February revolution in 1917[22], "the slogan was 'out of the imperialist way by overthrowing the bourgeoisie', and after the third and final revolution in October 1917, emphasis was on the restoration of economic welfare to develop the nationalised industry for the benefit of all and drawing into it the vast masses of peasantry."

Looking at the imperial Chinese society[23]...

> "Marxists and especially the Chinese communists, on the other hand, treat the imperial and even the Kuomintang period as a form of feudalism, meaning a society in which most of the land is owned by land-lords whose main income derives from rent."

In this form of domination over the peasantry, the dominating class rarely[24] "created an urban trading and manufacturing class", and, as a result, the link between the governing class and the rest of the population became superficial and weak and was "liable" to snap under any severe strain. The exploitation of peasantry through tenancy and that of the working class through imperial service in the initial stages, and in subsequent stages with starvation and sub-human life with superhuman labour, leaves no alternative but rebellion and insurrection. Frantz Fanon[25] argued that the "starving peasant outside the class-system is the first among the exploited to discover that only violence pays". The tying of peasant to the existing patterns to maintain the perpetual state of exploitation by formulating certain standards such as colorful ceremonies with great pomp and show for collective indoctrination, *i.e.*, to convince the landless peasantry and exploited proletariat to follow the traditional way of "respecting" their elders and thus continue slavery, is but like an organised crime, the systematic indoctrination; or in a stronger statement, it is a forced generation of flunkeyism in the mass media, which is in fact, fundamentally inexcusable. The exchange of labour through kinships of the nucleus family system was another significant feature to maintain the same perpetual state of exploitation of landless peasantry and working class. Social prestige and full-fledged membership in the society went automatically to those who possessed a piece of land. When all these grievances and the wounds of suffering of the oppressed can no longer be healed by "pedagogical" philosophies of the ruling oppressors, the latent anger and grievances of the oppressed simply burst out, oppression culminating in a rebellion. But[26] "the rebellion requires more than the destruction of the prevailing social bonds", because gangsters emerges as a result of this weakening of social law and order. Now, it is the quality and foresightedness of the leaders to organise all the dissatisfied (the rebellious peasantry, the angered proletariat and the upsurge of gangsters) in such a coherent way and in an integrated package that all these dissatisfied emerge as a single force and set the collective stage for a revolution. Such was the leadership of Mao Tse-Tung[27], that the—

> "Mass media of insurrection that began in 1927 (when Mao dissented with Chiang Kai Chek) and culminated in a Communist victory of 1949, was a land short of peasantry. And thus came a decaying of the Imperial China and an abrupt halt of all the "rights" and "privileges" of more than twenty-five foreign nations since the days of the Opium War 1939-42."

It was the "rights" and "privileges" of these foreign nations, which prompted Mao Tse-Tung to describe that China was worse than a colony.

When a poor peasant, in acquiring a piece of land, pays a relatively higher price than a large landowner, he is facing the prospects of a "superhuman labour and sub-human life".

This is primarily because he has to work hard, largely manually, to till the small field as compared to the large landowners and capitalistic farmers, who either mechanise the farm or hire the agricultural labourers. A landless peasant cannot maintain himself in either case, *i.e.* he cannot buy the large land, and, if he works as an agricultural labourer, he gets exploited and faces the prospects of sub-human life and super-human labour. Lenin had insisted that, in the eventual progress from individual to specialised agriculture, the small peasant must be persuaded, not coerced. But what we have observed, after more than 50 years since Independence?

As would be witnessed analytically in the following pages (Abolition of zamindari or formation of agrarian capitalism? the behaviour of the ruling elite and their alleged association with large landowners is[28]:

> "Reminiscent of the member of a group, who after every hold up hide their share in the loot from the other member who are accomplices and prudently start thinking about their retirement."

A detailed discussion of this particular picture, excluding land reforms, appears in Professor Gunnar Myrdal's Asian Drama, *An Enquiry into the Poverty of Nations* (New York; Pantheon, P., 1968, Vol. II, Chapter 21). We have reached the paradox in which if we are caught in a bribe and corruption case, we can get out by bribing.

Under such organised and systematic gangsterism, especially in a multiclass capitalists society (officially, however, there is a socialistic pattern of society in India), it is the landless peasant and the working class who pay the heavy price. These people are stagnating[29], deplorably in unbearable poverty. When they awaken to the unutterable treason of their leaders, their agitation well intensifies against the exploiting upper ruling class and may be waged all over India.

Except for the somewhat radical governments in the States of West Bengal and Kerala, there has been an absolute lack of political will to implement the land reforms, because of obvious vested interests. An editorial on 16-3-1992 in one of the most prestigious and widely circulated newspaper of India, the Hindustan Times[30], well summed up the necessity of land reforms for agricultural and industrial growth on the one hand, and the lack of political will in the country to implement land reforms on the other:

> "The countries that have succeeded in modernizing their economies all began with far-reaching land reform. In our country, the major obstacle in the way of meaningful land reform has been the lack of political will at the highest level."

To counter this sorry plight of land reforms in India during nearly the last five decades since Independence, V.P. Singh, on assuming the office of Prime Minister on 2.12.1989 pledged that[31] :

> "Land reforms is a very important area for the poor and ... that legislation regarding land reforms would be placed in the Ninth Schedule of the Constitution so that litigation is limited and undue money and time of the poor are not taken away and they really get justice."

But the government of V.P. Singh was short-lived, lasting less than one year. After Mr. V.P. Singh came Mr. Chandrashekhar whose government was also short-lived, less than a year.

P.V. Narasimha Rao, became the Prime Minister of India in June 1991. On 14th March, 1992, while inaugurating a Conference on Land Reforms which was attended by the Chief Ministers and Revenue Ministers of all the States of India, he clearly said that as far as the implementation of land reforms is concerned, something has gone "seriously wrong", adding further that:

> "It does not add to anyone's glory that very little has been done since Independence with respect to land reforms and land records despite assurances given by all the political parties during and after elections."

It is to be seen to what extent the pledge is translated into a positive action in favour of the poverty-stricken landless labourers of India, who have been living a sub-human life even after 50 years of Independence.

B. THE CASE OF THE STATE OF BIHAR

> *Bihar is a notorious State concerning land reforms. The atrocities being committed by landowners in Bihar on landless labourers, shall perhaps exceed even those committed by Hitler. .. Big farmers and landlords in the State of Bihar want even today (after 50 years of Independence) to enslave the landless labourers."*
>
> —*Editorial, Dainik Bhaskar, 1988*[32]

For the last 10 years Laloo Prasab Yadav has been running Bihar like fiefdom, untouched by elections and undeterred by rivals. For every allegation of misgovernance or corruption, he would have the same answer. Conspiracy by "Brahmanical elements, Manuvadis, Communal forces, or agents of western imperialism."

—*Editorial, Times of India, 2001*[33]

But there can be little doubt that decline has accelerated in the 12 years of Laloo raj. (His wife, Rabri Devi becoming chief minister is but a small part of the sorry saga.) All development activities gone to sleep under a leader who commanded enough political capital—perhaps as much as the state's first chief minister, Srikrishna Sinha—to take bold initiatives in agriculture, industry, infrastructure, health and education. Instead, lawlessness has become the law.

—*Editorial, Hindustan Times, 2002*[34]

Ranjeet Bhushan reported the grim picture of this State of Bihar as follows[35]:

Shri Laloo Prasad Yadav was Chief Minister for two terms during 1990-97. The then Governor of Bihar, Shri Sunder Singh Bhandari, reported to the Government of India that, "more than 51,000 people have been killed; more than 7,000 women were raped; there were close to 24,000 dacoities and more than 17,000 kidnappings. And these were the registered cases" (*Outlook Magazine*, New Delhi, October 5, 1998, p. 18). And Editorial in the *Indian Express*, New Delhi stated that to hold free and fair elections in Bihar is, "a challenge to the entire political system" (Bihar *vs.* Ballot, 14-9-1999, p. 8). Laloo Prasad Yadav resigned from the Chief Ministership due to charge sheet filed against him by the CBI on various corruption cases involving thousands of crores of rupees in many cases among them *chara* (Fodder) scandal. After his resignation, his literally semi-literate wife, was "Elected" by the MLA's of the Bihar Assembly as Chief Minister. Poll survey shows that 74 per cent want Laloo's wife Rabri Devi be sacked as Chief Minister.

"Grim Picture of Bihar: A State of Total Bankruptcy

- Total assets: Rs. 16,391 crore
- Total liabilities: Rs. 20,773 crore
- Excess withdrawals between 1990 and 1997: Rs. 4,000 crore
- Amount earmarked in Eighth Five-Year Plan: Rs. 13,000 crore. Amount spent: Rs. 5,000 crore
- Lowest growth rate in the country: 1.7 per cent
- Population below poverty line: 56 per cent
- More than 85 per cent of the state is without power
- Teachers, doctors and thousands of employees working in state owned undertakings have not been paid for a long time."

Rakesh Sinha while mentioning about the crisis of governance in Bihar wrote that:[36]

> Bihar is plausibly the only state which has failed to take advantage of economic reforms....There were 65.78 lakh families below the poverty line in 1993. In 1999, this went up to 93.99 lakh.
>
> However, the magnitude and frequency of the massacres have increased in the last 10 years. Since 1990, 93 incidents of mass killings have taken place . In 1999 more than 100 people lost their lives in six major massacres. This year, seven major incidents have already taken place, killing 86 people. The number of reported cases of kidnappings for ransom increased from 290 in 1993 to 346 in 1996 to 436 in 1999.

In this chapter, I would like to discuss the agrarian structure of the Government of India, specifically the new land reforms initiated and approved by the Parliament of India, after Independence in 1947. I will discuss this topic, using as a case study, the State of Bihar. In Bihar, an intensified agitation was waged under the leadership of Jaya Prakash Narayan, at the beginning of October 1974, and the State witnessed a three-day general strike, on November 4, 5, and 6, 1974.

In 1947, the Indian subcontinent was divided into India and Pakistan (now Pakistan into two: Pakistan and Bangladesh) on strict religious basis, Hindu and Muslim, The British thus laid down the seeds of hatred between the two nations. On January 26, 1950, India became a Republic. A written Constitution wad adopted with a parliamentary democratic system, the chief executive being the Prime Minister, the President being the symbolic Head of the Union. Soon after Independence, as a first step, a balance sheet was drawn to set up a list of priorities. Undoubtedly, India being an agrarian economy, strategies were formulated on land reforms. The first intensive land reforms came through the Acts of Parliament of India in 1951.

In this section, I propose to analyse what had been formulated since 1947, and what has actually been accomplished. While the virtues of a socialist pattern of society, the abolition of zamindari (a system instituted in India by the then British authority in 1793), were being preached, in fact (as we will see later), its turnout has been the formation of an agrarian capitalism. The exploitation of the landless peasantry and the agricultural proletariat persists. Stagnation in implementing the land reforms programmes passed by the Acts of Parliament continues.

The various measures taken up to change the inequitable land tenure system and to make the agricultural sector more progressive, can be broadly classified under five major groups[37]:

1. Abolition of intermediaries;
2. Measures for reducing rents, conferring on tenant's permanent rights over the land subject to the landlord's right to resume certain area for his personal cultivation, and enabling the tenants to acquire its ownership on payment for a reasonable price;
3. Fixation of ceilings on land holdings and distribution of surplus land;
4. Consolidation of holdings; and
5. Development of cooperative farming and large-scale cooperative village management.

The general strike and revolutionary ferment in Bihar was waged against the growing inequality between the various sections of society. A miserable and sub-human life, without any human dignity, was faced by the bulk of the population. On the other hand a handful of people the ruling class and elite, the bourgeoisie an association with large landowners, enjoyed affluence. The reason for the agitation in Bihar was simple. Bihar had virtually failed to implement any past of the agrarian reforms, passed by the Acts of Parliament and the State Legislative Assembly, as described below:

The geographic location of Bihar is one of the major causes for the appalling poverty of the landless peasantry and agricultural proletariat. It is located in the Gangetic region. Probably not a single year goes by, without a severe flood, human casualty, and intensive destruction of crops.

The Government of Bihar made the first post-Independence legislative attempt to abolish the Zamindari system by introducing the Bihar Abolition of Zamindari Bill, 1947. The bill was amended by the then Governor-General of India under the Government of India Act, 1935 (pertaining to provisions dealing with mineral rights in that State) and was approved on June 6, 1949 and published as Bihar Abolition of Zamindari Act, 1948. This was challenged in courts by the Zamindars and was repealed in 1949. On January 26, 1950, the Constitution of India was adopted. India became a republic and the post of Governor-General was abolished. The President of India, Dr. Rajendra Prasad, gave his assent to the Bill, which was passed as Bihar Land Reforms Act, 1950. But again, this new version of the Act was challenged in the High Court of Patna, as infringing Article 14 of the newly adopted Constitution of India, which read[38] "State shall not deny to any person equality before the law within the territory of India". The Court again ruled against the Government's decision and the Act was declared invalid. The Central Government then amended the Constitution, which validated this Act. But the desperate Zamindars then challenged the validity of the first amendment of the Constitution by the Parliament of India in the Supreme Court of India. The Supreme Court, however, in its verdict of

May 1952, upheld the validity of the first amendment of the Constitution.

Through this Act, the Government of Bihar, 'legally' abolished the "rights" of zamindars and tenant-holders to collect the rents. During this whole process of challenge and counter-challenge in the Courts, the land reforms, initiated in Bihar right after Independence, were dragged under the political rug, a result of the determined efforts of the zamindars to maintain their interests.

While battling on legal fronts, the zamindars were also preaching the virtues of 'peace' and to avert the political "civil war" in the country, which they threatened if their rights were seized by the Government. This is evident from a telegram, which the zamindars sent to the then President of India, Rajendra Prasad (who was a lawyer, a man of conservative orientation, and a representative of one of Bihar's leading caste, Kayastha. His general background was such that he was expected to give a sympathetic hearing to the plea of Zamindari interests), which read[39]

> "Hon'ble Rajendra Prasad. New Delhi, Honour Gandhi Jayanti (Birth Anniversary of Mahatma Gandhi) Appealing peace. Kindly drop abolition Zamindari. Save Country from Civil war."

What is abhorrent in this telegram is that the selfish interest of the zamindars, invested in them by the super-selfish interests of British colonists (East India Co., 1873 permanent settlement, under which the zamindars were authorised to collect rent and deposit 90 per cent of it in the account of East India Co., retaining the remaining 10 per cent for their services), continued to drag the issue of land reforms, and dragging with it the name of Mahatma Gandhi, President Rajendra Prasad replied to the telegram[40]: "I have never been able to understand the justice or fairness of depriving a man of the management of his property. ... I do not think there is any sanction in the Congress resolution for forcibly dispossessing people of their property before they have been compensated for it." This reply reflects that the Congress High Command was not in a mood to take any radical land reform action and implementation of whatever was proposed earlier.

Now let us look at what has been happening since 1952. First, there is the question of ceilings on how much land a person can hold under the Bihar Lands Reforms Act, 1950. The Bihar Agricultural Lands (Ceilings and Management) Bill, 1955, proposed: "Ceilings that were recommended varied from twenty-five acres for certain canal irrigated lands (Class I) to seventy-five for hilly, sandy, or other lands not yielding paddy or rabi crops, or any other crops such as chilies, sugarcane, or tobacco", and that[41] "under no conditions a landholder be allowed to hold land in excess of his personal ceiling." The meaning of economic holding and maximum ceiling were not defined anywhere and are thus ambiguous. Under First and Second Five Year Plans[42] "Ceiling is considered to be fixed as three times the size of family holding, but the definition of family holding is completely missing." Each individual State provided the missing element as they pleased. In Bihar, it varies from 30 acres of canal flow irrigated land to ninety acres of Class V (hilly or sandy) land.

The outcome of all these ambiguities, when discussed in the State legislature, brought so much friction within the ruling Congress party that is simply remained a bill and could not be enacted. The majority of the members in the ruling Congress themselves are large landowners. These conflicts culminated in the defeat of Mr. K.B. Sahay in 1957 general elections. This was the defeat of a person, who initiated and fought and who even counter-challenged the 1950 Act in the Supreme Court and finally had its validity upheld. But the struggle for implementing the agrarian reforms continued by the constant efforts of K.B. Sahay's group, which managed to remain in the Bihar Cabinet. As a result, in 1961, came an amendment, Bihar Act XII of 1962. The Class V land (sandy, hilly, etc.) was reduced to 60 acres from 75 acres. The loopholes and ambiguities in the Act, however, continued. More embarrassing is the fact that sugarcane farms were exempted from the ceiling provisions. Sugarcane is one of the major products of Bihar. The Bihar Act XII of 1962 provided loopholes that allowed the large landowners to abuse the ceiling limitations as they pleased. The Act permitted each landholder to transfer portions of his holding to relatives and non-relatives within twelve months following the commencement of the Act. During this twelve month period, large landowners could easily divide the land among their kins. The joint family system of India provided a sound infrastructure upon which to build the superstructure to beat the law.

Now let me return to the fundamental question, namely, the socialistic pattern of society as preached by the fathers of the nation, particularly Jawaharlal Nehru, when they adopted the constitution. As yet, there is no implementation of the interests of the landless peasantry and agricultural proletariat as mentioned in various Acts. All the discussion, so far, is on how to keep the land (by the ruling elite) in the name of "democracy" and "law and order".

It is this systematic and organised gangsterism that has led me to question whether Independence in India was a transfer of power to Indians for the alleviation of the downtrodden, landless peasantry and the exploited working class, or was it simply a change of masters. By the latter, I mean has a group of super elite replaced[43] by the outgoing colonialist English. It is evident that the ruling elite and in association with large landowners wear a white mask over their native skin, and continue the practices inherited from the colonial administration of Imperial Britain. While under colonial rule, the division was between the British masters and Indian coolies; after the Independence, it was transformed into 'master' and 'servants'. A few upper class families replace the English. The rest of the population had their status changed from coolies to "servants".

The Central Government, "confessing" the lack of implementation of the land reforms programmes, said "it has accordingly been decided to establish, for this purpose, a land reforms centre at Pune on the campus of the Gokhale Institute of Politics and Economics[44] where the main thrust will be on speedy and effective implementation of land reforms. The irony is that the programme that was initiated right after

Independence in 1947, 1951, and 1974, has come down to establishing a center in a university to "study" the problems of implementation. After 50 years of Independence, the actual implementation is yet to be seen. Following the 1970 land grab movement in Bihar; the Home Ministry of the Central Government recognised that the twenty years of planned rural development had not transformed the agrarian structure in much of India. Land reforms generally have not benefited the actual tillers. The unrest continued to build up and culminated in a land grab movement in August 1970. This was a militant and often violent agitation, led by Naxalites, the pro-Mao Communists of India. It is this continued exploitation that led Professor Januzi to conclude that[45]:

> "Events in Bihar, particularly as they are related to agrarian reforms that are promised and subsequently denied can be considered illustrative of processes going on in many regions of India in which those who are low in the in equalitarian, agrarian hierarchy have begun to be agents of change."

Bihar witnessed the nations first post-Independence agitation against the existing socio-economic inequalities and the constant effort by those in government and upper class strata to maintain the *status quo.*

The three-day general strike (November 4, 5, and 6, 1974) in the entire State spread to other parts of the country, including Delhi itself, the capital of the nation and eventually led to the declaration of national "emergency"[46] by Indira Gandhi to protect her family rule by absolutely unconstitutional measures.

Nehru's Agrarian Programme: An Out-and-out Failure

Jayaprakash Narayan and Jawaharlal Nehru and all other were freedom fighters under the leadership of Gandhi, Nehru, who became Prime Minister in 1947 did not like radical social changes in general and land reforms in particular. Vinoba Bhave and Jayaprakash Narayan, like other leaders emphasised that

India should not follow a revolutionary course. It was the naive belief that large landowners would come forward by themselves and give parts of "excess" land to Vinoba Bhave's Bhoodan Movement who, in turn, would distribute it among the landless peasantry. It was assumed that this goal could be accomplished without much class conflict.

But what was witnessed in Bihar was a large-scale of seating and filling of landless and particularly of the scheduled caste helpless peoples. An Editorial in *Dainik Bhaskar*[47] (Hindi) reported:

> "Bihar is a notorious state concerning land reforms. The atrocities being committed by landowners in Bihar on landless labourers, shall perhaps exceed even those committed by Hitler. ... Big farmers and landlords in Bihar want even today to enslave the landless labourers."

Hemendra Narayan summed-up the "progress" made in land reforms in Bihar as claimed by the Government[48]:

> "Given the track record of the Government on this count, the declaration of the Chief Minister (Mr Bhagwad Jha Azad on October 2, 1988) could be described only an empty rhetoric. Though it is true that Bihar has never lagged behind in legislation on land reforms, its achievement has been miserable, with the landed class dominating both the executive and political wings, thwarting the ceiling laws enacted in the State in 1961."

Jayaprakash, who waged and led the agitation in India in 1950's regarding social reforms, was the only Gandhian follower and Marxist intellectual. He initiated the Bhoodan Movement with vigor. But soon after, the hope that large landowners would donate their land, turned into disappointment and frustration. Jayaprakash Narayan[49] said on April 19, 1954: "I am ashamed we Bihar is took a vow of 3,200,000 acres in order to solve the land problem of our province. When we have not even enough workers to solve the land problem in Bihar, how can we ever solve it in the rest of India?

We can easily see that, on both the battlegrounds, legal and moral, the agrarian reform programmes have been an out-and-out failure. Both the central and provincial governments have confessed it. Professor Moore[50] rightly reached the conclusion that Nehru's agrarian reforms have been an "out-and-out failure". The proofs and explanation have been discussed in the preceding pages.

I must agree that agrarian reforms and implementation are a critical choice in the complexities of Indian society where kinship loyalties, caste system, religious faith (superstitious though it may be), still prevail. There is no single equation that could contain all these variables so as to select what form the present situation will eventually take and what the course of action should be. It is also hard to speculate how far the systematic denial of the rights of the poor the landless peasantry can go on.

The time has gone (probably never to return), when illusions of the new acts of parliament and of new promises could be cherished. The incidents in Bihar in 1970 and the intensified agitation of 1974 clearly indicate this. Similar agitations are going on in other parts of the country, which are not described in this book. The one action that can possibly avert these tragic incidents is an effective and immediate implementation of what has already been proposed. Certain social, political, and economical reforms, especially those dealing with land reforms and agrarian structure, are now virtually inevitable.

The landless peasantry is now asking the bourgeois coalition of the ruling class and the large landowners[51]. "What is the meaning of your fine political arrangements when the rich can still oppress the poor"? Until now, in India, peasant discontent has not yet taken any significant revolutionary ferment and has been mainly confined to Gandhi's philosophy of fraternity, again a return to an idealised village community. In my opinion, the upper class of Indian society cremated all the ideals of Gandhi along with him. I am aware of the seriousness of what I am writing, but this is a fact. This is what the country is witnessing. This is a national tragedy.

What is appalling is that no systematic attempt was made by either the State Governments or the Central Government to inform the rural areas that agrarian reforms were being instituted and that the abolition of zamindari was underway. This information was given and when some political campaign was undertaken. As a result, this information was greeted with growing frustration and disillusionment. Government-controlled communications media never really provided intensive information enabling the landless peasantry to fully understand the relevant mechanics of agrarian reforms. Solidarity in such cases emerges, mixed with latent anger, develops a rebellion psychology, and enhances the revolutionary ferment.

Letting the peasants realise their grievances, is not sufficient for creating revolutionary ferment. There must be complete integration of all the circumstances and various socio-economic and political factors, and careful consideration of each action or in-action. If the revolutionary leaders do not consider such factors, we may reach the paradox[52]. "What does modernisation mean to peasantry, when sooner or later they are its victims." A tragic situation arises when; the ruling bourgeoisie crushes the revolutionary ferment. This is particularly true of the Indian peasantry, for it is tied to the prevailing system and kinship loyalties. Under such conditions, possibilities for insurrection are only present, only to be uprooted. Professor Moore seems right that:[53]

> "Hinduism might go a long way towards explaining the pasivity of the Indian society. Eventual success or failure of the revolutionary ferment will largely depend on how much and, in what way, these existing social institutions are taken into account not to mention the passivity of the Hindu thought belief."

This is largely so, because agriculture is being practiced for centuries on indigenous methods bullocks, the plough, and depending on the mercy or havoc monsoon rains. The government has never attempted to practice the principles of socialism as they have been preached. But[54] "there has been considerable talk about socialism, enough to disturb business community quite seriously; in actuality very little has been done". The atmosphere of action "has become the substitution for action". In response to such an assessment, concluded Moore, the western liberal observer almost automatically responds that even if Indian agrarian policy, indeed her economic policy as a whole, has been rather long on talk and quite short on accomplishments, at least there has not been the brutality of communist modernisation. Some sacrifices in speed, the argument runs, are necessary for the sake of democracy.

This ideology against the communist model of modernizing the society fact, poses a fundamental question to my mind, or I suppose if I can say so, to the thought of every sensible student of political economy, and that is; of what use is democracy or communism or socialism or dictatorship or any "ism" or any ideology to a person who simply has nothing to do and nothing to look forward to in his or her life? None is unaware of the ever-increasing wretchedness in rural areas[55]. In such a situation, it is very hard to make a distinction between the "brutality of communist modernisation" and the "necessity of sacrifices for the sake of restoring the ideals of democracy."[56] It is at this stage that revolutionary ferment dominates the mass media. In Fanon words[57], "the starving peasant, outside the class system, is the first among the exploited to discover that only violence pays. For him, there is no compromise, no possible coming to terms".

"Rural revolution achieved, largely by peaceful means, would in any case deserve a place in history."[58] This is the very basis of the Gandhian concept of decentralised development. Unfortunately, I do not see any part of the Gandhian philosophy being translated into practice. The chain of great social reform, as envisaged in Gandhi's idealist view of life (Ram Rajya), are remote possibilities, for the whole structure of the Government of India, particularly under the prime ministerships of Indira Gandhi and Rajiv Gandhi since 1967 and 1984, respectively, both at the central as well as the provincial levels, was largely based on sycophancy, corruption, nepotism, and mismanagement.

The more liberal views such as Vinoba Bhave's Bhoodan Movement or Gandhi's stress for a place for Harijans (children of God, untouchables) in society have proved fine in theory only and never really been practiced in society.

As I discussed earlier, under the topic "Abolition of Zamindari or formation of agrarian capitalism", the reluctance of the national bourgeoisie for an effective implementation of land reforms programmes passed by Parliament, reflects their selfish interest in association with the ruling upper class, and they continue to enjoy their monopoly over landownership. The social conditions of the masses of the Indian subcontinent have not much improved since Independence; the increase in social welfare and other such activities is next to negligible. The mass celebrations of public holiday such as on Independence Day (15th of August) or on the show of military muscle and folk culture such as on the Republic Day (26th of January) are not likely to pursuable the ordinary poor poverty-stricken fellows in the rural areas or the unemployed in urban centres. In the words of Frantz Fanon[59]:

> "The peasant who goes on scratching out a living from the soil and the unemployed man who never finds employment do not manage, in spite of public holidays and flags, new and brightly coloured though they may be, to convince themselves that anything has really changed in their lives."

The overall faith of the landless peasantry and that of the exploited working class in the promises of politicians is now weakening, as is obvious from the agitation in the country in general, and in Bihar in particular. This deteriorating situation cannot be controlled by policy or by the military for a long time. Those who are at the bottom of the society[60] "Where the general rule is that the great wealth is surrounded by the greatest poverty", have begun to be the agents of change, and therefore are a potential source of challenge against the existing inequality and exploitation. This has been discussed in the beginning of this section, 'Structure of peasantry and class analysis'. We notice that, while there is overwhelming labour done by the family in tilling their rather small piece of land (the middle peasant) there is formation of agrarian capitalism, as discussed in the previous topics. The latter is a result of the exploitation of landless peasantry by the large landowners and capitalist farmers. This is what Jayaprakash Narayan voiced in Bihar in his struggle against the then prime minister Indira Gandhi, and, as Moore said, the structure of peasantry may[61] "became the seed of reactionary anti-capitalist sentiment in the countryside".

There has been little rather virtually no success in eliminating the structure of peasantry since Independence. The structural of the Congress is primarily based upon large landowners and big industrials that the government does not want and cannot challenge for its own survival. The same trend continued even when famine was waging in the country in early 1970 and thousands were dying. An officer from Bengal[62] said: "There is a lot of brave talk about food procurement, a lot of radical rhetoric, but the State Government (West Bengal) and New Delhi do not want to upset the politically powerful landowners."

In the existing structure of peasantry, the landless peasants find themselves unable to organise for a number of reasons—there is no such thing among them as class solidarity and they are scattered. Hamza Alavi[63] said:

> "They stand divided among themselves by their allegiance to their factions led by their masters. Political initiatives thus rest with the faction leaders who are owners of land and have power and prestige in the village society."

This unorganised and agricultural proletariat and unemployed workers are useful to their masters in two ways[64]: (1) subjective factors such as "*their backward mentality. ...*" and

(2) objective factors such as *"their dependence on their masters for their continued livelihood."*

The commercial and industrial classes form hardly any alliance with the downtrodden peasant, the exploited proletariat, and even the discontented intelligentsia. During the beginning of the Japanese aggression of China in 1937, all the landed upper class, the industrialists and the commercial classes of Imperial China, joined with the foreign imperialist interests and scrambled into large urban areas to protect their property and wealth. Mao Tse-Tung and other revolutionary leaders "availed" themselves of this opportunity and managed to organise the ignored and isolated landless peasantry and led a successful revolution in 1949. This communist victory in China changed the face of that country once and for all. In 1950, the victorious Mao said, *"we are a blank cheque on a bankrupt bank, on which we could write anything and build any structure."* Today, in 1990s, we see that there is in China a sound infrastructure for its socio-economic development and political stability. China, the most populous nation on earth, has emerged from the grip of poverty and illiteracy and all foreign interests. At present, she stands with pride, as the richest, militarily powerful, and politically stable nation among the so-called Third World Nations.

More or less, the same was the case in Russia on the eve of the Revolution of 1917[65]:

> "The commercial and industrial classes were not a suitable ally for the angry peasants. The Russian bourgeoisie was much smaller and weaker in the country as a whole. ... The Russian bourgeoisie was tied by many strings to the Tsarist Government. ... The demands of the Russian peasant were brutally simple: to get rid of the landlords, divide up the land, and of course, stop the war."

While the Tsarist Government had refused any concession and the peasants were the main victims of slaughter and were paying the price with their lives, the Bolsheviks led by Lenin, agreed to meet the peasants' demands to seize the power.

After the successful revolution in Russia, there was a similar passion in Indian leaders. The socialist group of the Indian National Congress wanted to form a similar kind of strategy. Gandhi believed that private ownership was not good. He was against the use of violence against strike the large landowner. Gandhi said[66] that *"it is my firm conviction that, if the State suppressed capitalism by violence, it will be caught in the evils of violence itself and fail to develop non-violence at any time"*. It will be worthwhile, however, to note that Gandhi said those words when the whole of India was under the British. In my view, Gandhi probably feared that any violent action against Indian large landowners would force them to seek the shelter of the British. The British could then use these dissented landowners against the Indians, and then could further entrench their policy of "divide and rule" (between Hindu and Muslim).

Britain has long been driven away but her legacy—the colonial symbol in the form of dependency and cultural schizophrenia[67] still persists. In other words, the national bourgeoisie and ruling elite and upper class people have alienated themselves from those whom they rule. They have put on a British mask over their brown skins.[68] The main beneficiaries of the fruit of Independence have been approximately 10 per cent of the population. Although the agitation in India may not find allies among the upper class strata, it may draw upon the vast majority of the middle-class, urbanised, educated, young, and unemployed.

In the absence of implementation of stipulated land reforms programmes, the development of agriculture (on which would depends the process of agricultural mechanisation and rural industrialisation[69]) is bound to remain a failure. Programmes for the development of agriculture that the Government of India announces largely remain on paper[70]. The agricultural development programmes of the Government are just too many. If even half of them could be sincerely implemented, India has the potential (both in natural[71] and in human[72] resources), to be a nation once again what it once was before its systematic destruction by the British began—*"The Golden*

Bird". An Editorial in the prestigious newspaper of India, Hindustan Times summarised the political wrong doings in Bihar as follows:[73]

> But there can be little doubt that decline has accelerated in the 12 years of Laloo raj. (His wife, Rabri Devi becoming chief minister is but a small part of the sorry saga.) All development activity gone to sleep under a leader who commanded enough political capital-perhaps as much as the state's first chief minister, Srikrishna Sinha-to take bold initiatives in agriculture, industry, infrastructure, health and education. Instead, lawlessness has become the law.

NOTES

1. Editorial, *The Hindustan Times*, New Delhi, 16-3-1992; p. 13.
2. National Institute of Rural Development, India Rural Development Report, 1999, Hyderabad, India, pp. 19-20.
3. *Ibid.*, pp. 19, 32.
4. Karl Marx, *Das Capital*, Volume I (Moscow: Progress Pub., 1954); p. 337.
5. Mahatma Gandhi, *Collected Works*, Vol., 64 (New Delhi: Ministry of Information and Broadcasting, 1976); p. 73.
6. Jawaharlal Nehru, *Discovery of India* (London: Marion Books, 1946); p. 535.
7. *India Year Book*, reference manual, 1977-78, Publications Division, Government of India, New Delhi, 1978; p. 216.
8. Charan Singh, *Economic Nightmare of India* (New Delhi: National Pub., 1981); p. 147

   Responding to his critics, Ch. Charan Singh claimed the following regarding his being a Kulak, or a friend of the big or rich landholders:

   "Being a public man and having been criticised by my political opponents for being a "Kulak" or a friend of the big or rich landholders, perhaps, it should not be considered self-adulatory on my part if I state here that every term, idea or concept incorporated in the revolutionary land reform legislation of Uttar Pradesh was my contribution. Many a measure in this connection met with a stiff opposition from some of my colleagues in the State Cabinet. I held charge of the Revenue portfolio, whether as a Parliamentary Secretary, Secretary, or a full-fledged Minister from April 1946 to

April 1959 except for two brief periods in 1947-48 and 1951-52. When I resigned from the State Cabinet in 1959, the portfolio was made over to a colleague who was virtually opposed to abolition of landlordism, had no love or sympathy for the poor and the under-privileged and entertained no anxious moments if the latter were ejected from the land under their plough." p. 125.

9. Jawaharlal Nehru, *op. cit.*, p. 535.
10. The beginning of socio-economic reform movement in 1974 in Bihar, and its scattered fall-outs in the rest of the country that threatened the position of the then Prime Minister Indira Gandhi, culminating in the declaration of the so-called "National Emergency" that lasted till March 1977, when India Gandhi's party was virtually routed out of power in the national general elections.

    During this period of "National Emergency", Mrs. Indira Gandhi ruled through the most repressive and measures; all the opposition leaders were put behind the bars, except for the sychophants whose venality was unquestionable.
11. *See* Section B of this Chapter for the case study of Bihar.
12. Frantz Fanon, *The Wretched of the Earth* (New York: Grove Press, 1961); p. 169.
13. *Ibid.*, p. 61.
14. K. Goush and H.P. Sharma (Ed.), *Imperialism and Revolution in South Asia* (New York: Monthly Review Press, 1973); p. 295.
15. David Mitrany, *Marx Against Peasant* (Chaptel Hill: The University of North Carolina Press, 1951); p. 25.
16. V.I. Lenin, *The Land Question and the Fight for the Freedom* (Moscow: Progress Pub., 1965); p. 140.
17. David Mitrary, *op. cit.*, p. 71.
18. *Ibid.*, p. 195.
19. By the term "new nations" what I mean are the nations who have emerged free from the British and French and other Western powers' colonial yoke. For example, India was free from the British colonialism in 1947; Algeria from the French in 1963, etc.
20. Joseph Stalin, "*Foundations of Leninism*" in A.P. Mendall, (Ed.), Essential Works of Marxism, p. 248.
21. *Ibid.*, p. 248.
22. *Ibid.*, p. 254.
23. B. Moore, Jr., *Social Origins of Democracy and Dictatorship* (Boston: The Beacon Press, 1966); p. 163.
24. *Ibid.*, p. 205.
25. F. Fanon, *op. cit.*, p. 61.
26. B. Moore, *op. cit.*, p. 214.
27. *Ibid.*

28. F. Fanon, *op. cit.*, p. 173.
29. *Ibid*, p. 167.
30. *The Hindustan Times*, New Delhi; 19.8.1988; p. 11.
31. *Employment News Weekly*, New Delhi; Dec., 23-29, 1989; p. 1.
32. Editorial in *Dainik Bhaskar* (Hindi daily newspaper), Bhopal; June 27, 1988; p. 4.
33. Editorial, Siege Within, *Times of India*, New Delhi, 8-3-2001, p. 8.
34. Editorial, From bad to worse in Lalooland, *Hindustan Times*, New Delhi, 19-8-2002, p. 8.
35. Ranjit Bhushan, Suspended Anination:The Battle for Bihar, *Outlook Magazine*, New Delhi, October 5, 1998, p. 16.
36. Rakesh Sinha, Crisis of Governance (case of Bihar), *Economic Times*, New Delhi, 7-11-2000.
37. S.C. Jha, *A Critical Analysis of Indian Land Reform Studies* (Bombay: Asian Studies Press, 1971); p. 1.
38. F.T. Januzi, *Agrarian Crisis in India: The Case of Bihar* (Austin: The Uni. of Texas Press, 1974).
39. *Ibid.*, p. 14.
40. Part of a quote in *Ibid.*, p. 14.
41. Report of the select committee on Bihar Agricultural Lands (Ceiling and management), Bill I, clause 6, 1955, Quoted in F.T. Januzi, *Ibid.*, p. 73.
42. S.C. Jha, *op. cit.*, p. 5.
43. See Chapter 6, "Poverty *versus* Polity in India and the Third World".
44. Fourth Five Year Plan (1969-74), Planning Commission Government of India, New Delhi; 1974; p. 183.
45. F.T. Januzi, *op. cit.*, p. X.
46. The beginning of socio-economic reform movement in 1974 in Bihar, and its scattered fall-outs in the rest of the country that threatened the position of the then Prime Minister Indira Gandhi, culminating in the declaration of the so-called "National Emergency" that lasted till March 1977, when India Gandhi's party was virtually routed out of power in the national general elections.

    During this period of "National Emergency", Mrs. Indira Gandhi ruled through the most repressive and measures; all the opposition leaders were put behind the bars, except for the sychophants whose venality was unquestionable.
47. *Dainik Bhaskar* (Hindi Daily), Bhopal, 27-6-1988; p. 4 (translated by the Author from the original in Hindi).
48. *Indian Express*, New Delhi, 13-10-1988; p. 8.
49. Quoted in F.T. Januzi, *op. cit.*, p. 99.
50. B. Moore, *op. cit.*, p. 495.
51. *Ibid*, p. 495.

52. *Ibid.*, p. 455.
53. *Ibid.*, p. 455.
54. *Ibid.*, pp. 389, 407.
55. For details, see Chapter 1.
56. B. Moore, *op. cit.*, p. 407.
57. F. Fanon, *op. cit.*, p. 61.
58. D. Mitrany, *op. cit.*, p. 2.
59. F. Fanon, *op. cit.*, p. 169.
60. *Ibid.*, p. 172.
61. B. Moore, *op. cit.*, p. 477.
62. *New York Times*, Oct., 27, 1974; p. 28.
63. Humza Alavi, "Peasant and Revolution", in K. Gough and H.P. Sharma, *op. cit.*, p. 332.
64. *Ibid.*, pp. 332-33.
65. B. Moore, *op. cit.*, p. 481.
66. Quoted in B. Moore, *Ibid.*, p. 375.
67. For detailed analysis, *see* Sures Chandra Jain, *Education for Socio-Economic Development*, Concept Pub. Co., New Delhi, 2004, especially the chapter on Inferiority complex: the servile imitation of the colonial rulers of the West.
68. See Chapter 6 of this book, "Poverty *versus* Polity in India and the Third World".
69. As described in Chapter 2 of this book.
70. Details see Section B of Chapter 6 of this book. Some of there programmes are as follows:

    I.R.D.P. Multiple cropping, high-yield varieties, soil and water conservation, fertilisers, quality seeds, plant protection, agri-industries corporation, agro service centres, drought-prone area programmes, dry-land development, rainfed problem development, farmer's service societies., etc.
71. See Chapter 5, "Utilisation of Natural Resources."
72. See Chapter 4, "Mobilisation of Human Resources."
73. Editorial, From bad to worse in Lalooland, *Hindustan Times*, New Delhi, 19-8-2002, p. 8.

4

Mobilisation of Human Resource

"The development of a country is brought about by people, not by money. Money, and the wealth it represents, is the result not the basis of development."

—Julius K. Nyerere[1], 1967

"To be effective..., a development strategy must reflect population concerns among its primary objectives. Similarly, a population strategy must reflect development concerns. It must link population programmes to programmes in health, education, housing and employment, among others. Indeed, it is only through such linkages that sustained and sustainable development can be achieved".

—World Forum on Population[2], 1989

A. THE DEMOGRAPHIC SETTING

The population in South Asia and share of population of South Asia in world population are given in Table 4.1 and Table 4.2 respectively.

The demographic background of India from 1901 to 1991 is given in Table 4.3, along with decadal growth, average annual exponential growth rate and progressive growth rate. Table 4.4 gives distribution of population, sex ratio, density, and growth rate of population from 1911 as compared with 1991.

Table 4.1: Population in South Asian Countries and in Various Regions of the World for 2000

Region	*Population 1999 in millions*
	p. 279
India	997.5
Pakistan	134.8
Bangladesh	127.7
Nepal	23.4
Sri Lanka	19.0
Bhutan	0.782
Maldives	0.278
South Asia	**1329**
World	5975
Low Income	2417
Middle Income	2667
Low and Middle Income	5084
East Asia & Pacific	1837
Europe and Central Asia	475
Latin America & Caribbean	509
Middle East & North Africa	291
Sub-Saharan Africa	642
High Income	891

Source: The World Bank, World Development Report 2000/2001, Oxford University Press, New York, 2000.

Table 4.2: Share of South Asia in World Population

Region of the World	*1995 (Actual) (in %)*	*2025 (Projected) (in %)*
South Asia	22	24
East Asia (excluding China) and the Pacific	11	12
Americas	13	13
China	21	18
Arab States	5	6
Europe and CIS	18	12
Sub-Saharan Africa	10	15

Source: Mahbub ul Haq and Khadija Haq, *Human Development in South Asia 1998*, The Human Development Centre, Karachi, Pakistan (Oxford University Press, Karachi, Pakistan), 1998, p. 10.

Table 4.3: Population of India: 1901-1991

Year	*Population*
1901	238,396,327
1911	252,093,390
1921	251,321,213
1931	278,977,238
1941	318,660,580
1951	361,088,090
1961	439,234,771
1971	548,159,652
1981	683,329,097
1991	843,930,861

Source: Data condensed from Census of India, 1991.

Table 4.4: Distribution of Population, Sex Ratio, Density and Growth Rate of Population

India/ State Union Territory	*Population 1991*			*Sex Ratio*		*Growth Rate*	
	Persons	*Males*	*Females*	*1981*	*1991*	*1971-81*	*1981-91*
(1)	(2)	(3)	(4)	(5)	(6)	(7)	(8)
India	843,930,861	437,597,929	406,332,932	934	929	+ 24.66	+ 23.50

Source: Data condensed from Census of India, 1991.

Table 4.5 gives data on the progress made in literacy from 1971 to 1991. Table 4.6 gives age distribution of population, 1981. Table 4.7 gives major religious populations.

Table 4.5: Progress of Literacy 1971-1991

	Literacy Rate (%)			*Change in percentage points*	
	1971	*1981*	*1991*	*1981 over 1971*	*1991 over 1981*
Total	29.48	36.23	42.94	6.75	6.71
Males	39.52	46.89	52.68	7.37	5.79
Females	18.70	24.82	32.52	6.12	7.70

Source: Data condensed from Census of India, 1991.

Table 4.6: Age Distribution of Population 1981

Age-group (years)	*Population (Mn)*			*Percentage distribution*		
	Male	*Female*	*Total*	*Male*	*Female*	*Total*
0-14	42.4	41.3	83.8	12.3	12.9	12.6
5-14	93.4	85.9	179.3	27.1	26.7	26.9
15-19	33.9	30.1	64.0	9.9	9.4	9.6
20-24	29.0	28.3	57.3	8.4	8.8	8.6
25-34	47.3	45.8	93.2	13.8	14.3	14.0
35-44	38.0	35.1	73.1	11.0	10.9	11.0
45-54	29.2	25.5	54.7	8.5	7.9	8.2
55-59	8.5	7.9	16.4	2.5	2.5	2.5
60 & above	22.0	21.2	43.2	6.4	6.6	6.5
Total	343.9	321.4	665.3	100.0	100.0	100.0

Source: *Statistical Outline of India*, 1989-90, Tata Services Ltd, Bombay, 1989; p. 40.

Table 4.7: Major Religious Populations

1971 1981[1+2]				
Religious Community	*Population*	*Percentage to Total Population*	*Population*	*Percentage to Total Population*
Hindus	45,34,36,630	82.72	54,97,24,717	82.63
Muslims	6,14,18,269	11.20	7,55,71,514	11.36
Christians	1,42,25,045	2.59	1,61,74,498	2.43
Sikhs	1,03,78,891	1.89	1,30,78,146	1.96
Buddhists	38,74,942	0.71	47,19,900	0.71
Jains[3]	26,04,837	0.48	31,92,572	0.48
Other Persuasions	21,84,955	0.40	27,66,285	0.42
Religion not stated	36,083	0.01	60,217	0.01

1. 1981 figures are based on the information on religion of heads of households tabulated through the household schedule.
2. Excludes Assam.
3. Other persuasions total of remaining religions as given by the respondents.

Source: India (Year Book), 1987, Ministry of I&B, New Delhi; p. 14.

According to 1981 Census, there were 3,949 towns in India, 5,57,137 inhabited villages[3] and 48,087 uninhabited villages.

The population of India in 1991 was 844 million. The

present growth rate of population is approximately 2.1 per cent. This means that approximately 33 babies every day or 14.25 lakh every month or 17 million every year. It is the addition of 17 million every year that causes concern for already severely hard-pressed socio-economic system to provide for food, housing, education, health care, and employment. If we compare the population of India with the total population of the world, we notice that India has about 2.4 per cent of geographic area of the world, with a population of nearly one-fifth of the world population.

At the current rate of population growth, we add approximately 17 million people every year, while the total population of Australia itself was approximately 16 million in 1987. So we are adding the equivalent of the entire population of Australia every year!

Table 4.8 gives population growth from 1965 to 1969 and projections up to the year 2000 country group-wise.

The population of the world in 1840 was just 1000 million or 100 crores. It was approximately 5 billion in 1987. It is expected to be approximately 6 billion by 2000.

Table 4.8 shows that the population of low developed regions is twice as great as that of developed nations. If this trend between populations of the low developed regions and developed regions continues, as it has up until now, the future prospects will certainly be:

More Developed Regions

The level of living could possibly be equal or even below that of the under-developed nations of today. The bare necessities of life (health, education, and other essential services) will be desperately and hopelessly overstretched.

Low Developed Regions

Concentrated efforts are being made today against protection from hazards of climate People are born, live, and die in the streets. If population projections prove true, we will have two options:

1. Either some natural havoc or catastrophe will occur before these projections come true. Then the anticipated gloomy, and frightening situation is solved by itself.

   Or

2. If not, we would then be working to keep lessening mass starvation and widespread malnutrition.

Table 4.8: Population Growth and Projections

	Population (million)		
	1989	*2000a*	*2025a*
Low-income economies	2948	3633	5201
China and India	1946	2300	2950
Other low-income	1002	1333	2251
Middle-income economies	1104	1354	1954
Lower-middle-income	682	842	1224
Upper-middle-income	423	512	730
Low-and middle-income	4053	4987	7155
Sub-Saharan Africa	480	679	1311
East Asia	1552	1822	2307
South Asia	1131	1396	1959
Europe, M. East, & N. Africa	433	533	813
Latin America & Caribbean	421	513	700
Severely indebted	554	673	926
High-income economies	830	884	965
OECD members	773	813	862
+ Other	58	71	103
Other economies	323	349	404
World	5206	373	692
Oil exporters (excl. USSR)	265	373	692

- Low-income economies are those with a GNP per capital of \$ 580 or less in 1989.
- Middle-income economies are those with a GNP per capita of more than \$ 580 but less than \$ 6,000 in 1989. A further division, at GNP per capita of \$ 2,335 in 1989, is made between lower middle-income and upper-middle-income economies.
- High-income economies are those with a GNP per capita of \$ 6,000 or more in 1989.

Source: Data condensed from World Development Report, 1991, The World Bank, Washington, DC, 1991; pp. 254-55, p. X.

This continued population growth in the low developed nations, without an equal economic growth, will cause economic backwardness and will help maintain the continuing state of parasitism with frustrated desires. Has not the "world gone mad" and "lost control of its demographic growth as Rene Dumont[4], a noted French philosopher remarked?

B. POPULATION PROBLEM OR THE PROBLEMS OF THE PEOPLE

In Arusha Declaration[5], the then President J.K. Nyerere of Tanzania made the following remark the utilisation of human resources:

> "The development of a country is brought about by people, not by money. Money, and the wealth it represents, is the result and not the basis of development".

Most of the population of the abundantly populated developing countries such as India dwells predominantly in rural areas. The ignorance and illiteracy of the villagers remain major hindrances in the implementation of many development programmes. The elimination of illiteracy and the spread of family planning are bound to make the over-population emerge as an asset in development planning, which otherwise remains a liability.

So what is needed is an approach to solve the problem of ignorance and illiteracy. Mahatma Gandhi[6] said that, if we could remove the poverty and illiteracy of the Indian villager, we will "have the finest specimen of what a cultured, cultivated, free citizen should be".

However, the policies of the government are directed at solving the problem of population, without first attacking the problems of the people, *i.e.*, illiteracy and poverty. Unless we solve these 'problems of the people", the 'population problem' can at best scratched at the surface only. So, is there 'population problem' or the 'problem of the people' ? In the case of India, it is the latter that feeds the former, while both are serious problems and analysed in the text.

The policy planners in India have[7] so thoroughly been influenced by the Western concept of population policies and problems that they ignore the fact that the Western policies do not apply to countries like India or even China. Mao Tse-Tung, while refuting "the absurd arguments of Western bourgeois economists like Malthus that increases in food production cannot keep pace with increases in population", said that the solution to the big population problem "is production".

Investments in education and economic development are distinct entities. This is particularly true of under-developed nations, especially in the initial stages of development. Investment in education is both a means of production and an item of consumption.

The expression "means of production" is used in the sense that it creates an attitude and environment to accept the challenge of change. It develops the self-image and personality of the individual. It enables the individual to realise and recognise human dignity. It promotes self-discipline and widens the horizon. It produces the necessary skills upon which rests the basic infrastructure for economic growth. In a broad sense, investments in education broaden the people's mind, which, in turn, creates the advanced means of production, though indirectly. However, this non-measurable and indirect yield on educational investment largely depends upon entrepreneurship and administrative organisation of the country's bureaucratic system and administrative structure. A better organised society and maximal use of talents and training obviously results in a comparably higher yield than in an inadequately organised society. With organised society, I take into consideration various other factors, such as the geographical dispersion of populations and the discontinuity in the learning process. This is particularly true of locations where enough facilities are not available for education, forcing a person to discontinue studies. Investment in education in the underdeveloped nations should be viewed in its entirety as a means of development upon which an overall superstructure of the nation's existing social-cultural-economic value system develops, and thus provides the invisible but sound foundations upon which the human resources can be mobilised.

Investment in production is an item of consumption in the sense that it is independent of its possible application to the means of production. This is so, because it has an indirect return and non-measurable considerations. The development of infrastructure of the existing social-cultural-economic value system makes a direct contribution to improve an overall image of the level of living, though largely non-materialistic. Thus[8] an "initial heavy investment in human resources development is necessary to get a country started on the road to self-sustained growth." Why National Planning?: In the developing areas of the world, the quantitative and qualitative aspects of educational planning should be considered in accordance with a systematic mixture of classical and contemporary values and a foresighted and forward-looking approach. This is imperative if we have to transform the existing educational system (which has its roots deep in colonialism) into one that best suits the needs of the nation and meets the needs of the people. The heavy influence of the foreign experts' "advice", has often been far away from the realities of the host nation. In India, a successful attempt is being made to switch over the communication media and the language of instruction in the colleges and universities from the English language (imposed by the colonial Britain) to regional languages. No doubt, a person can better communicate, read, and write in his/her mother tongue. The conclusion of Professor Don Adams[9] that "at the higher education level, the decline in the use of English has threatened to lower the standards of even the most distinguished institutions," as far as I can see, is away from any realities and foresightedness. There is no single developed nation in the world, which has a foreign language as communication media in its schools, but there are many under-developed nations, which still continue the foreign language as the medium of instruction in their schools and even feel "proud" of maintaining the colonial symbol. This argument of professor Adams holds truth, if any, only in the transitional period, the time that is required to transform the colonial system into the nations own value system. In the long run, however, this argument has no justification and holds no truth whatsoever. The guidelines used in educational planning must be based on the socio-cultural values of the nation. This is imperative if we are to avoid the feeling of frustration among the masses alienated in their homelands because of the language media in the schools. This is what we see today in India.

An Indian is more "Americanised' than an American in the United States, an Indian is more 'French' than the French in France, and more 'English' than the English in the U.K., and so on. This transformation is particularly true of civil servants and bureaucrats. This is also the case with almost all the rest of the underdeveloped world, with a few exceptions (notably the China during the rule of Mao Tse-Tung or the example of Julius Nyerere in Tanzania). Mahatma Gandhi[10] wrote in 1937:

> "Babubhai is M.A., LL.B., but this does not mean that he remembers what he reads of politics. And he is not the only one. We, all of us, educated people, have proved ineffective in giving political education to the people. Our English education has made us useless. The heritage left by our forefathers has been squandered."

The developing nations cannot develop unless attempt are made to maintain their own identities and to restore their self-images based upon their own cultural value systems. This approach, and recognition of this philosophy, in my view, is one of the foremost determinants of the development system of the under-developed nations. Professor Rene Dumont[11] rightly concluded,

> "The backward and dominated countries, first of all, must rid themselves of our economic and cultural influence, which saddles them with a civilisation model hindering their authentic development,...permanent education must be promoted for all, beginning with functional literacy...this would lead to true development..."

This input into the development system can only be through educational reforms at the national level. The output of this will then be what Nyerere envisaged in "Education for Self-Reliance."[12] The title itself is sufficient to explain what I mean. To achieve self-reliance through education.[13] "Indeed, in many cases the goal of national prestige may be given higher priority than economics and social progress." This can only be achieved when the existing educational systems of the poor nations (which can never put the national theories into practice because they are alienated) are shattered and new ones built courageously, thoughtfully, and carefully. This change will encourage the people to re-unite with their own way of life and the realities from which they now have been divorced. The immediate consequences of the new approach will not only result in bridging the ideological differences among the nations but also the bureaucracy in the developing nations will function more effectively because of the superstructure (of administration) which would have been re-shaped and streamlined to meet the needs of the people.

The development of economy, which necessitates a geometrical progression in the means of production, cannot develop without quality in skills, entrepreneurship, and management capabilities. Thus, we have to apply the scientific methods dialectically to study the prospective educational planning and to preview its far-reaching economic and social consequences. The planned social and economic development of a country can only fit into its general development planning when there is a balanced development of the whole educational system, qualitatively and quantitatively. Explicitly manipulated, strategical concepts in educational planning (such as formal and non-formal) and recommendations of International Conferences and Congresses) will undoubtedly accelerate the implementation process. The assistance and know-how from the centrally planned economies (socialist countries) has a substantial role, as opposed to that from western countries. Capitalist nations do not practice planned economics at home and no longer teach economic planning sciences in their school. But under-developed nations can only develop through a planned economy. A planned economy, in its practical form, means a socialist pattern of society. In other words, assuming this statement as the base, the under developed nations have to build a socialist educational system. This is exactly what was done in China during the Cultural Revolution under Mao Tse-Tung, especially in the areas of general economic disciplines. After the change of government in 1949,[14]

> "The new socialist oriented society quickly made schooling available "for all"—from part-time literacy classes for the uneducated through all the academic levels, whereas before 1949 education in terms of schooling was limited to a tiny section of population—the ruling elite of China's semi-feudal and semi-colonial society."

This socialist strategy helped the Chinese Government to remove disparity between highly educated and less educated. In the national modernisation process, all the participants, whether highly talented people and statesman/women or those from the labour force are engaged in productive activities. The works of artists, musicians, theologians, writers, and philosophers (all those engaged in development of socio-cultural economic value systems) are all equally important. At the national level of development, the active cooperation of all participating members from every sector of life is necessary and should be the integral part of the system as a whole. This is like an automobile, which can only move smoothly when both the axles and both sides of wheels on each axle move together. And of course, only one pair of wheels on either of the axle drives (the one that provides power, transmitted by the engine), and the other pair of wheels on the other axle is driven (the one that gets push from the driver).

The General Conference of UNESCO[15] at its 11th Session (Paris, 1960) approved the establishment of a Regional Centre for the training of educational planners, administrators, and supervisors, which was organised in New Delhi. Such a regional organisation is of paramount interest to the participating under-developed member States because, by and large, they all face the same problem. Thus we can conclude that the use of education for economic development requires[16], "(1) a national policy that recognises that education is a requirement for economic development, and (2) consideration of the specific characteristics[17] education must posses to facilitate such development."

The utilisation and integration of the human resources with the nations' socio-economic development policies, is one of the major tasks facing the abundantly populated under-developed nations. A strong grip and a cohesiveness between the two is not only desirable but also imperative if we have to transform Gunnar Myrdal's[18] "Poverty of Nations" into Adam Smith's "Wealth of Nations". The likely consequences of maximum utilisation of raw human material can be viewed as economic growth and bridging the gap between the rich and the poor as equitably as possible. We have to review the basic question of national development not only in terms of growth in GNP but also in terms of effective utilisation of raw human material in more productive activities. This means we need to draw a visible line that clearly distinguishes between labour force and material goods. And further, it has to be made clear that labour should not only include those engaged in productive activities but also those engaged in social and cultural modernisation processes such as, artists, musicians, writers, philosophers, theologians, and others. The interdependence of raw human material and economic growth is clearly reflected when we hear,[19]

> "The plaintive cry of the villager about his son's future, 'without an education how can he escape my miserable fate, judgments of the consumers of education in the developing countries about the economic value of the education are worth listening to".

If the educational planners realise this "cry", which is legitimate and farsighted, there arises the very complicated problem of what kind and what quality of education the son will have. The answer lies in the nation's existing socio-cultural value systems. Educational planners have to envisage how to coordinate the "status" of the son who has been educated with the rest of the society to which he belongs. In other words, "balance of brain" has to be maintained within the society among the high-talented people—those engaged in productive activities and those who not only keep alive the nations' value systems but also mould it for the future generation outcome. By the latter I mean authors, philosophers, statesmen, artists, musicians, etc.

In the developing nations, there is widespread reluctance on the part of the professional to move closer to the rural areas, where they are badly needed. Educational planners can discourage this reluctance by measuring the output in rural areas as substantially higher than in urban areas in terms of monetary values. To discourage this reluctance[20], *"appeals, compulsion, and exhortations are not the practical aspects of this problem"*, as the noted Indian scholar, Dr. V.K.R.V. Rao remarked.

Encouragement of discovery and subsequent cultivation of potential talent from the human raw human material must be the motto behind educational policies, in addition to the general educational development. To achieve this goal, it is imperative that attention be focussed at the very roots. Who is to discover the potential talent? To a certain degree, it is the teacher. It means, in human resources planning, that the role of teachers' colleges must have a significant place. To enhance this, the best brains and intellectuals must be encouraged to enter colleges and universities or related careers. This is a common practice in the developed nations. But largely this is not happening in the developing nations. Professor Butts of Columbia University Teachers College points out that[21]:

> "All over the world teachers are universally poorly paid and often times poorly prepared. Despite its newly won recognition as a key to development, education continues to come off second best in manpower allocations in comparison with more lucrative, the more powerful, or the more glamorous occupations."

As a matter of fact it is the teacher who shows the path of light to the coming generations. What was said in Bhagwad Gita so many thousands years ago holds true today with even greater force[22]:

> "Whatsoever a great man does, the other men also do (imitate); whatever he sets up as the standard, that the world (people) follow."

It was truly said in a UNESCO document in 1970[23] that the:

> "Planners have to given an absolute priority to teacher training, and, before that, the selection of those who train the teachers. That is the key to the whole system, and a mistake at this point jeopardises the entire operation."

The high quality of teachers is an absolute necessity if we are to get the maximum output from the massive investments in educational research and training and make an intimate relationship between educational methodology and teaching materials. The requirements of best intellectuals in teaching fields[24], "is almost as true today as it was in the old days of the Indian Gurukul[25] system of education that the teacher is the central figure in the educational system." It seems clear that general national development depends upon the success of the educational system. Hence the quality of the teacher is the central focus in the overall development planning, though indirect. G. Myrdal[26] also remarked "regardless of difficulties, a higher social, economic and professional status for teachers in primary schools are an absolute precondition for raising the levels of teaching."

For the national development approach to be realistic (not merely an idealistic), the population-abundant under-developed nations have to utilise their abundant masses largely unutilised to their full potential, into the development system as an asset which stands now as liability. This means that interdependence has to be formed between economic growth and raw human material. The under-developed so far have not been able to utilise their masses to play a positive role in the growth process, (except, mainly in China). We need the growth process that[27] "will release mass energy, stimulate mass participation, and motivate mass involvement," as Professor V.K.R.V. Rao said. The Soviets, before the Revolution in 1917, had the same problem under Charism heavy drag of mass poverty. The infrastructure after the Revolution was designed on cooperative and collective[28] bases that mobilised the masses successfully. The Chinese has adopted the same pattern since 1949. The carefully coordinated participation and involvement of the idle masses in the development process will help bolster the development of the psychology[29] of self-help dignity, respect, and reliance as envisaged by Julius Nyerere. The creation of this psychology and its consequent changes in the attitude of the idle masses will also help remove existing social inequalities.

In November 1989, a Conference was held at Amsterdam, by the United Nations Population Fund in cooperation with the Government of Netherlands. This Conference was called The World Forum on Population for 21st Century. This Forum made a number of declarations. One of the declaration is to link the population control strategy with the socio-economic development strategy itself[30]—

> "Recognising that the implications of the sheer magnitude of such numbers are staggering, the forum has set before itself a development strategy resting on several pillars. Strengthening political commitment, acceleration and broad basing of resource mobilisation, strengthening the role and status of women; strengthening family planning, maternal and child health programmes; heightening community awareness; expansion of national programmes and seeking great international cooperation in the field of population activities, with specific reference to socio-cultural conditions. The accent on increased women's role in decision-making and management of population policies is particularly relevant for developing countries and

significantly forms the crux of India's family planning programme."

The development of skills increases the capacity of performance and productivity of the labour force, and there is no limit to the development of skills. A proper forecast of skilled manpower, consistent with growing industrial needs and maximally utilizing the labour force, is highly desirable for a realistic approach.

The main objectives of the under-developed nations' educational system must be to make a clear interdependence among education, social factors, and economic considerations. Educational planning in the social factors must be capable of creating the feelings of realisation and recognition of "self-image". Human dignity and the recovery of national conscience among the long oppressed masses, both from colonial oppression and the inferiority complexes built-in due to terrible poverty is something higher than the mere struggle for goods, literacy, clothing, and protection from the hazards of climate. Most of the under-developed nations are young and getting rid of the colonial yoke. They should no longer retain their former colonial rulers' system. The work of churches, and especially of the missionaries, in developing areas in the educational fields must cease, because they have been working since the colonial period in co-operation with the colonial rulers. Their trends, attitudes, and curricula has hardly been adapted to the needs and value systems of the poor nations in which they function.

The next objective must be the promotion of economic development, with a thoughtful and careful coordination among the scientific, technical, and skilled personnel. There must be a government decree to abolish any form of discrimination on grounds of race, colour, sex, language, religion, political and/or other opinion, social and/or national origin, etc., and be observed effectively. Education must be the birth right of every individual in accordance with the UN Declaration of Human Rights. There must be active participation of women in educational planning and development, especially at kindergarten and primary level. The guiding principles for building a learning system would be (1) comparative advantages, (2) multipurpose service, (3) decentralised planning, (4) rational allocation of funding for all constituent programmes.

Integration between formal and non-formal education is imperative so that only the best meets the individual nation's needs within her own guidelines. It is essential to measure the dimension of problems—of poverty, of unemployment, and of under-utilisation of human resources. This dialectical approach will maximally utilise the mass energy, released due to careful planning, into the national economic growth. Decentralised planning and subsequent rational allocation of funding should be such that learning services such as agricultural and handicraft skills are available locally in town or village areas. This is because practical oriented programmes carry relatively higher importance in teaching adults. Before the formulation of a plan the individual opinions of leaders in their respective fields must be taken into serious consideration of sociologists, of economists, the corporate opinions, public surveys, of thinkers and philosophers, the international recommendations especially those of the various UN specialised agencies (notably UNESCO) in the field of education. Conclusions of various International Congresses and public opinion through polls, comprising people from all spheres of society, should be consulted. Commissions of enquiry, which should be set up, should deal with many controversial matters and delicate issues. These opinions will help the planners to integrate organic and functional coordination in different categories—adult education and rural development; formal education; formal and vocational planning; scientific and technical policy formulation, etc. With the integration at these various levels[31], *"internal coordination between educational services themselves and between them and the other State services, all levels of public administration will be achieved with a far greater efficiency."*

The determination of objectives in educational planning is a complex activity, composed of a number of factors, most of them being autonomous in their respective functioning. However, all these various factors at work, when formulated with the national economic planning, are inter-related and interdependent elements. Here it will be necessary to distinguish between short, medium, and long-range objectives. While short-term plans form routine business and put checks and controls on the long-range planning, it is the long-range planning that provides the foundation stone upon which the short and medium-range planning function. While the long-term objectives determine the course of action, modes of development and various determinants for short-term planning, in turn, constitute the routine operational decisions. In China, during the first two decades after 1949, institutional and instructional reforms towards greater efficiency included[32]:

1. Elimination of all private institutions;
2. Control of enrolments by the State;
3. Realignments of colleges and departments in order to create long-range institutions of similar functions;
4. Establishment of a large number specialised institutes (rather than creation of new institutions); and
5. Creation of "specialists" to replace departments as the academic unit for instructions.

In this perspective, it is an absolute necessity that determination of objectives (in formulating the overall national educational planning) achieve an effective balance not only between quality and quantity of required manpower but also an integration of education with general development planning.[33]

Fig. 4.1, Components of Educational Planning Process[34], outlines an effective planning that can forecast the quality and quantity of manpower requirements based upon reliable sources. This must take into consideration the various related factors at work (see below) as a necessary integration of educational planning with general development of:

1. Agricultural development plans,
2. Economic development plans,
3. Forecast of manpower needs, and
4. Formulation of nations' scientific and technological policies.

In addition to this, it is essential to deal with the evaluation of educational plans, modernizing the educational management, such as planning priority options and programming the particular phases of general educational planning[35]; involving community in educational planning such as education of women, rural education for the sheer conservatives and superstitious and special education, for the deaf, blind, physically handicapped, and mentally retarded; strengthening educational planning by incorporation of all forms of non-formal education, and cultural activities such as educational museums and works of contemporary and classical arts, improving communication media by the establishment of theatres, music, and cultural centres, and providing State subsidizing of books and libraries (largely mobile type to reach the rural communities). In integration of researches of educational, economic, social, cultural, demographic, technical and scientific, regional, national, and international organisations, and further liaison with research in other fields such as collection of data needed for planning, diagnostic surveys, and case studies, etc., is also essential. Formulation of strategies as to what to choose amo006Eg other available alternatives, formal and non-formal education and subsequent forecasting of manpower requirements, functional literacy and rural development, training of teachers, and before that those who train teachers, are the other concepts of educational planning to be integrated with nations' general development planning. All these various details and sub-details have been sketched out in Fig. 4.1.

Thus a careful determination of short-range and long-term objectives in educational planning, based upon all the factors and various phenomenon mentioned in this topic[36], "may be the flower of economic development, but is also the seed" that

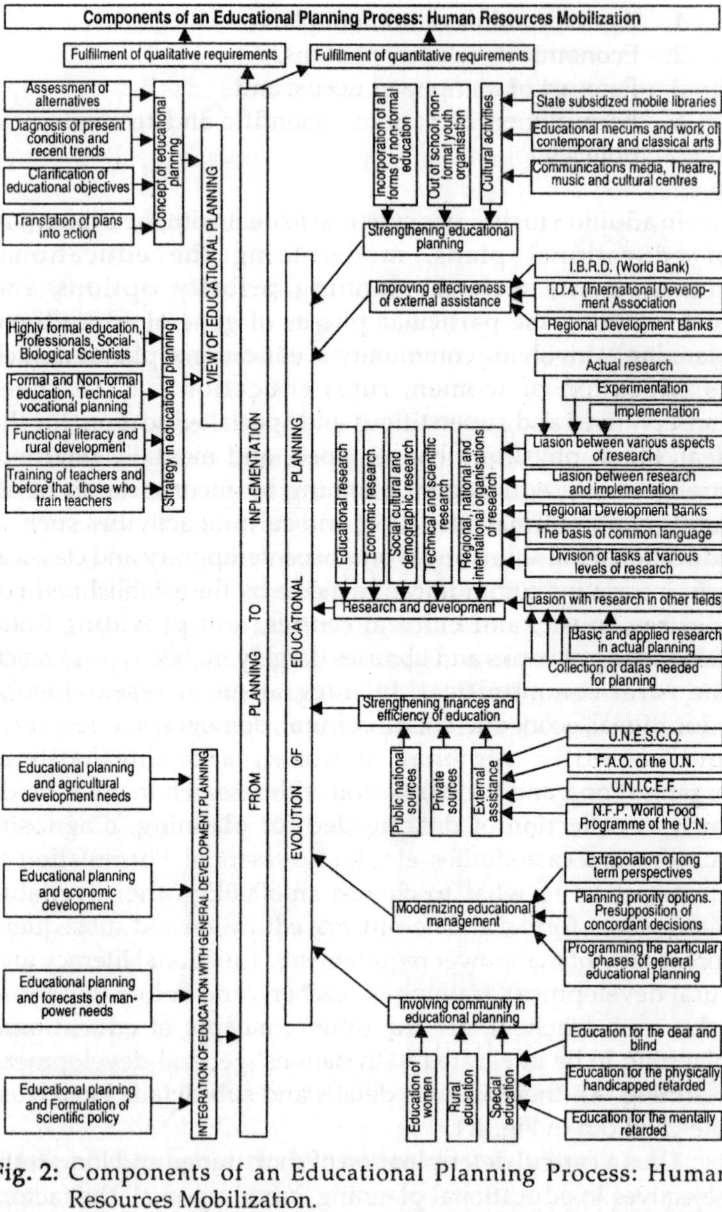

Fig. 2: Components of an Educational Planning Process: Human Resources Mobilization.

This diagram and these interrelated and interdependent elements are based upon author's interpretation of *Educational Planning: A World Survey of Problems and Prospects*, UNESCO, Paris, France, 1970.

may change the existing inadequacies and mobilise the raw human material through educational reforms. Finally, the most important aspect of any educational planning, must be such that it strongly and effectively emphasises the country's own socio-cultural values and commitments to the traditional heritage.

Having described the need for improving the quality of existing human resources, we now attempt to outline population problem and its control.

The object here is to solve the problems of the people, in so far as the existing stock is concerned. All attempts will be small to control the growth rate, making it to Z.P.G. (Zero Population Growth) to negative growth. It is in this sense that an improvement in the quality of human resources is essential, for then and only then the abundant population will be an asset in the process of socio-economic development against what is now a clear liability. So the problem of population is to be attacked in two district ways: First to increase the production capacities[37], qualitatively and quantitatively, and secondly, control on the population growth itself.

C. THE NECESSITY FOR POPULATION CONTROL

"Effective family planning can only be achieved if the Government of India, State Governments and political parties, irrespective of their ideological predilections combine to evolve a national consensus on giving top priority to family planning. Unless this is done, all talk of removing poverty will remain totally unrealistic, and in the larger perspective, our democracy itself may well get disrupted by the tidal wave of population growth."

—*Dr. Karan Singh*[38]*, 1988*

The necessity for population control; food production

On September 20, 1972, Dr. Kurt Waldheim, then Secretary General of the U.N. proclaimed the year 1974 as world population year, and a World Population Conference was held in August of 1974 in Bucharest, Romania. The U.N. - sponsored population and development talks indicate the human awareness about the population problem, the compound interest of human race. The population problem has a special significance among all the other problems[39] in the sense that it is directly human made and human controlled, if concrete efforts are made. The population problem is most acute and most dangerous in Asia. India alone has more than the combined populations of the continents of Africa and Latin America put together.

To understand the dialectics of population dilemma, a clearly defined distinction between population problem and problems of the people have to be analysed. It is, in fact, the problems of the people that create the population problem in poverty stricken nations. It is the inadequacy of the social welfare services in economically backward nations that causes the feeling of insecurity among the people. This is particularly true of poverty-stricken and usually illiterate masses. The desire to have more sons thus becomes an "insurance" in "too-old-to work" stages of life. The logic behind this reasoning is simple: survival is essential.

Improved hygiene, medical and sanitation conditions have lowered the death rate, making this problem of population growth even more acute. For example, an anti-malaria campaign in Mauritius after the Second World War, has cut the death rate to half in six years; the same thing happened in Sri Lanka in eight years. In India, where the death rate formerly ranged in the thirties, is now about nineteen per thousand.

Social and cultural values in many under-developed nations (for example, India and China) are such that a great prestige is attached to child bearing. Probably these societies feared that their generation would vanish. In a caste-system (notably India) and tribal societies (almost the whole of the African continent), it is not uncommon to see a minority group attaching itself to a majority group under the impression that birth control is aimed at them. These minority groups perceive population growth, not as a problem, but as genocide of their race. In this context, the responsibility of the majority groups to curb the population growth becomes even more vehement.

The superstitious, outdated, and mythical religious theories, which have continued from the time immemorial are a major hindrance. An attempt is made by leaders to keep these theories alive, without modification in the light of the present social requirements. They do not consider the future catastrophes that humanity eventually will have to suffer, if the present trend of population growth continues. The religious leaders tend to minimize, and even ignore the practice of birth control. Mohan Dharia[40] has well summed up the reasons of the gross failure of family planning programmes in India:

> "However, there are far more basic reasons behind this failure:
>
> 1. Our religious customs, traditions and preaching discard the very concept of family planning and they are not yet prepared for a new change.
> 2. Majority of the masses honestly feel that the birth of a child is a gift of God and to do anything against the natural course is to displease the Almighty.
> 3. Male superiority and importance of a son yet enjoy the same dominance in almost every family.
> 4. Poverty stricken people feel that additional hands are a source of additional income.
> 5. The death rate which was expected to come down to 10 per thousand by the year 1975, stood at 14 per thousand till 1981. There is so far no substantial reduction. In a country where 114 children out of every 1000 die immediately after birth, it is too difficult to motivate the people to adopt family planning.
> 6. To educate the people and impart knowledge about family planning is treated as something obscene and efforts are made to keep the youngsters miles away from this vital knowledge.
> 7. The political leadership in the country lacks the determination and courage in implementing this national programme of the most vital importance.

The greed for power has made them the slaves of voters."

Achieving economic and social development will be the most effectual contraceptive. It is obvious that all the direct family planning programmes have met with failure in almost all the under-developed nations. These programmes, however, have been quite successful in developed nations.

The Government of China realised, after 1949, that problems of the people must first be understood.[41] They

> "Carried through fundamental reforms leading to a rise in living standards, increasing emancipation of women, remarkably efficient sanitary infrastructure, and that by this means she has lowered her birth rate far beyond any other large country in the Third World."

In 1966, the UN Fund for Population Activities (UNFPA) was established to advise and give technical assistance to abundantly populated nations, at the request of the country concerned. Though the U.N. has had a Population Division from the start to study demography and to keep up-to-date population statistics, it was the creation of UNFPA in 1966, the Declaration of Population Strategy for Development in 1972[42], and the Proclamation of 1974 as World Population Year, highlighted by a World Population Conference[43] held in August, 1974, in Bucharest, Romania, which indicated the concentrated efforts taken by the World community.

The Tokyo Declaration clearly emphasised the problems of the people in resolving the population problem[44]:

> "While population has a direct effect on economic and social development and the human environment, conversely, policies in the fields of education, health, housing, social security, employment and agriculture have an impact on population and, therefore, require integrated national planning and coordinating action at the highest government level."

The World Forum[45] on Population for the 21st Century was

held at Amsterdam in November 1989. The United Nations Population Fund in cooperation with the Government of Netherlands organised it. This Forum clearly emphasised the need for linking population strategy with development strategy:

> "To be effective, a development strategy must reflect population concerns among its primary objectives. Similarly, a population strategy must reflect development concerns. It must link population programmes to programmes in health, education, housing and employment, among others. Indeed, it is only through such linkages that sustained and sustainable development can be achieved."

The Population Council and other organisations arrived at the same conclusion. Besides recommending the channelling of more funds into local development work, the relationship between population and development[46] was also stressed.

It seems highly unlikely that a direct approach to family planning (in forms of contraceptives, birth control pills, or now widely practiced abortion) can have any significant success. According to Pierre Praderand, one of the main reasons for failures of family planning programmes, notably in India[47], was that:

> "The approach advocated by the majority of experts during the sixties including myself was an almost total failure, and for a very simple reason: we concentrated on the provision of contraceptive services, practically without worrying about motivation. But contraceptive is a dynamic development."

The discussion thus far, proving an inverse degree of correlation between population-growth and development, is visible in Table 4.9. When we compare the annual rate of natural increase and GNP per capita between the developed and the underdeveloped countries we find that the higher the per capita GNP, the lower the birth rate.

Table 4.9: Annual Rate of Natural Increase in Population 1970 and GNP per Capita by Regions 1970

	Population per cent Rate of Growth	*GNP/ Per Capita US*
World	2	880
Low Developed Regions		
Africa	2.7	180
Asia	2.3	125
Latin America	2.9	560
Developed Regions		
North America	1.00	4660
U.S.S.R.	1.00	1790
Europe	0.80	2575

Sources:

1. Data condensed from A. Angelopolous, The World and the Rich Countries (New York, Praeger Pub, 1972) p. 52, Table 10.
2. *Ibid.*, p. 214, Table 25 (data condensed).
3. Data condensed from Trends in Developing Countries, World Bank, Washington, D.C. 1973, Table 2.5.

Therefore it can be concluded that problems of the people, i.e. the problem of economic underdevelopment are variable determinants, which could resolve the population problem. It is not direct birth control planning, which is needed. The government now[48], "needs to be able to base their actions on an adequate understanding of the trends, the instruments available, and the place of population policies within overall development policies".

So unless we do so, the only alternative left to us would be to cope with the widespread malnutrition and mass starvation. World War III would certainly "solve" the population problem, most likely forever, for there would then probably be another evolution of life. Dr. Karan Singh[49] warns:

> "...Effective family planning can only be achieved if the Government of India, State Governments and political parties, irrespective of their ideological predilections, combine to evolve a national consensus on giving top priority to family planning. Unless this is done, all talk of removing poverty will remain totally unrealistic, and in

the large perspective, our democracy itself may well get disrupted by the tidal wave of population growth."

One of the reasons for food shortage and population growth is the migration from rural to urban areas. Throughout history, there has been a migration from rural to urban areas. Various tables in this Section (A) show that there has been a continuous increase in the percentage of urban population both in the developed and developing nations.

In rural areas, agriculture is the main profession, and this migration may be the decisive factor in decreasing food production. The reasons for this migration may be social and cultural activities, faster means of communications and education facilities which are more available in urban than in rural areas.

The implementation of development projects by setting up of gigantic industrial plants to change the population distribution and, in the long-run, demographic characteristics of the people so affected, is one alternative.

Urban-oriented migration in developed and under-developed nations has to be viewed from entirely different angles. While "Western development models suggest that urbanisation is considered to be synonymous with modernisation, industrialisation and economic development"[50], in low developed nations, urbanisation usually does not provide large-scale job opportunities to migrants. These insufficient job opportunities are "particularly true in the case of the major colonial trading centres, such as Calcutta, Jakarta, and pre-revolutionary Shanghai."[51]

We are now faced with a critical situation, particularly in the developing regions, to obtain a higher level of food production with less people directly engaged in agricultural production.

To achieve this goal, we have to improve means of production and develop new ones, and to mechanise the traditional way of farming to keep pace with the growing population.[52]

Now, in the context, it would be wise to look at the food

situation today. It is clear that the low developed nations cannot feed their populations today. The question now arises that, when they cannot feed today, how can they hope to cope with that of today's double? Do we hope that they can produce enough food to meet the increased demand? What will be the extent of misery and frustrations in the year 2000? This is a challenge to every sensible person on earth today. What steps should be taken to cut down the present rate of growth and, at the same time, to double the amount of food production. Now let us be optimistic for a while and say that we develop tropical Africa for agriculture, develop new means of agricultural production, train more people, spend more money on agricultural research and development and if successful, celebrate the 'greenness' of 'green revolution'. Will we be able to double the food production by the year 2000? Then what? Do we have to struggle so hard and work so laboriously only to keep the present poverty and frustration and still see our[53] "five people mostly children, dying of starvation, as we are reading these lines"?

Higher yields and discovery of new means of food production are urgently required to keep pace with the geometric growth of population. Unless some drastic measures are taken to remove the non-linearity between food production and population growth[54], "we could only delay, not avert the ultimate collapse" as Rene Dumont observed.

Second only to the most dangerous and devastating problem facing the mankind today, the nuclear confrontation, is the population growth and environment deterioration. Population curbs therefore are a necessity for tomorrow for it would not be possible to provide all the material for consumption needed for an affluent life. The population, as I pointed out earlier, is expected to be double in 35 years based upon 1970 estimates. If we allow this population to grow at the present rate, we would need double the amount of food production. Unfortunately, population growth is like compounded interest of the human race, while material production in arithmetic ratio. Earth's resources are definitely not infinite. The total surface of the earth is a little less than

2000 million square miles. Approximately, three quarters is water and the remaining one-quarter is land. Out of this one quarter of land, about half is either too cold or too dry to cultivate. An additional one quarter of the remainder is similarly not available for cultivation for reasons of infertile soil, mountains or rocky terrains, excessive rainfall, or land required to build cities, roads, industrial estates, etc. The total land out of the whole earth available for agricultural purposes is approximately one-sixteenth.

The great oceans might swallow part of the land, but this available land area is not going to increase is any event. So the only solution to this problem is to curb the population growth. Population curbs are again necessary due to the disparity between death rate and birth rate caused by improved medical, sanitation, and hygienic facilities. They have drastically lowered the death rate and have tremendously increased life expectancy. Methods of immunisation discovered by Louis Pasteur, inoculation, X-ray, detection of tuberculosis, discovery of penicillin, etc. are but a few that save millions of lives.

Pressures of population sometimes even force rulers to wage war, to seize the territory and resources of other nations. Tragically, the starvation created by the waging of wars, kills more people than war itself. It is because a nation, under threat of attack or waging war, loses her means of consumer production for they are transformed into war-material production.

The unemployment rate will be at a staggering height—approximately 300 million more people will be added in the world labour force, calculated at the present rate growth two per cent world average. Out of these, about 230 million people will be in the underdeveloped nations. This is a serious threat to the social order increase in slum dwellers, poverty, under-development, and the environment population caused by it.

It is not that there is an inverse degree of correlation between underdevelopment and environmental deterioration. In industrially developed nations, emission of pollutants have caused environmental deterioration (such as accumulation of carbon dioxide which reduces solar radiation reaching the earth, thus cooling off the earth's surface; carbon monoxide produces incomplete combustion in auto engines; phosphates found in sewage and in detergents lead in marine deposits and its contents (fish and drinking water), oil origination in the operation of off-shore tankers. Radioactivity, created by nuclear power plants operation, creates genetic mutation, etc.) The underdeveloped nations, on the other hand, suffer from the pollution of poverty living in slums where as micro-organisms are caused by dissemination, lack of sewage, and other such disposal conveniences.

Thus, many environmental side effects are due both to the development of agriculture, industrialisation, transportation, etc., and to poverty, as mentioned earlier. The integration of economic development planning with inevitable environmental deterioration must be such that it can maximally minimize these side effects.

No matter what measure we take now, it does not seem likely that we can avert the doubling of this population at the end of this century. However, it cannot be argued that it is the population growth alone, which is responsible for the present socio-economic inequities. But it is, no doubt, the foremost contributing variable between or among all the other inter-related elements.

The discussion so far indicates a future catastrophe, which was predicted for the first time by Malthus[55], that "human population cannot grow indefinitely in geometric progression", Professor Rene Dumont cautioned[56], "if the population growth does not decline more rapidly, and if at the same time consumer habits and income distribution systems are not changed, total collapse will undoubtedly be inevitable." What Prof. Dumont forecast in 1974, Malthus[57] had already warned in 1796 that at one blow all the fond hopes of a harmonious universe will perish. What he offered in place of progress was a prospect dreary and chilling.

I can now make two alarming conclusions:

1. The developed nations are capable of blowing up this earth, if they arrive at confrontation with their nuclear machinery; and

2. The under-developed nations can well blow this earth by their "population bomb" the gloomy dimension of their misery and frustrations.

Prof. Angelopoulos[58] hypothesized it the other way round:

"At the present rate of growth, some 70 to 80 million persons annually will swell the number of people to be fed, while the world will have reached the limited of resource capacity. Then famine and misery will definitely check the present population explosion."

It is now therefore imperative, not for academic debates alone, that concrete and concerted efforts be made to defuse the population bomb, before it goes off. The Commission on International Development[59] concluded that *"there can be no serious social and economic planning unless the ominous implications of uncontrolled population growth are understood and acted upon."*

It is obvious now why it is necessary to curb population growth, but the question is how and with what drastic measures and what radical reforms.

One of the foremost variable determinants, as discussed in the beginning of this chapter, would be to coordinate the population problem with the problems of the people.

The Chinese, after 1949, initiated radical reforms in their traditional way of life and in their existing socio-cultural value systems. They have achieved successful results in birth control, which can be possible among all the other under-developed nations, but this has largely been possible by their labour-intensive planning[60]. In other words, the approach has been made directly, not on the family planning techniques, but on the economic development and elabourating various social services. The labour-intensive variable inputs have been the prime determinants in gigantic cultivation and irrigational projects. This has significantly curbed the population growth.

Japan is also heading successfully towards equalising death and birth rates. Industrial manufacturing plants, which provide job opportunities for all, have made it possible for her to maintain a constant population by the year 1990.

James Kocher[61] concluded the same thing, "I am aware of no evidence of sustained fertility decline—either spontaneous or induced through a family planning programme—ever having taken place in the absence of significant socio-economic development and modernisation."

However, in addition to this indirect birth control planning, the policy to curb population and the development policy are one and the same thing. There are several other variables, which, nonetheless, are of equal significance.

A clear distinction and analysis has to be carried out between scientific factors based upon logic and rationale and outdated irrational and traditional moral values. In dealing with the population problem, the law of logic and rationale sometimes does not work among the illiterates and superstitious masses. Today we need to develop new psychological methods to deal in such conservative cases so that they can be modified to fit into the present-day system. This is again essential, because society is changing with innovations in science and technology. To persuade these people, life cannot be explained only in its material aspects but also on cultural values.

Some of them may be in the form given below:

1. Literacy for adults.
2. Prevention of child labour and successful implementation of schooling at least until ten years of age.
3. Participation of women in decision-making and encouragement for jobs in traditionally male oriented societies. The right of women to decide the number of children and spacing.
4. Birth control planning and education about advantages of small families as part of high school curricula'.
5. Formation of "Zero Population Growth" (Z.P.G.)[62] groups should be the goal of every nation and should be coordinated and implemented through United Nations population division.

It seems rational that an integrated approach to population

control would be to combine the direct and indirect approaches, as discussed so far. This is exactly what was said in the first paragraph of the Tokyo Declaration[63]:

> "While population has a direct effect on economic and social development and the human environment; conversely, policies in the fields of education, health, housing, social security, employment and agriculture, have an impact on population, and, therefore require integrated national planning and coordinating action at the highest government level."

In the early 1990s, the money and material being spent on armament averaged about U.S. $ 1,000 billion a year.[64] If even a part of this money is converted into development efforts, it will have a tremendous impact upon the economic growth of the underdeveloped nations. This is again necessary, because a hungry person, a starving nation, can even upset peace any time. This is a serious threat to international peace and security, and adversely affects international relations. In realistic terms, this threat is even more dangerous than nuclear confrontation.

NOTES

1. Julius K. Nyerere, *Freedom and Socialism* (Dar-es-Salaam: Oxford University Press, 1968); p. 243.
2. *Economic Times*, New Delhi; 3.12.1989; p. 4 World Forum on population for the 21st century, held at Amsterdam in November 1989. It was organised by the United Nations Population Fund in cooperation with the Government of Netherlands.
3. An urban area is (a) all statutory towns, *i.e.* all places with a municipal corporation, municipal board, contonment board or notified town area, etc. (b) All other places which have: (i) a minimum population of 5,000; (ii) at least 75 per cent of male working population engaged in non-agricultural pursuits; and (iii) a density of population of at least 400 persons per sq.km. The inhabited and uninhabited villages stand for their literal meaning. *Source: India 1987* (Yearly reference manual, Ministry of I&B), New Delhi, 1988; p. 15.

4. Rene Dumont, "A World Gone Mad", in CERES FAO of the UN, Rome, Italy; November-December 1973; p. 17.
5. Julius K. Nyerere, *Freedom and Socialism* (Dar es Salaam; Oxford University Press, 1968); p. 243.

   This passage is from the famous Arusha Declaration. The Arusha Declaration marked a turning point in Tanzanian politics. The ideology of the country was made explicit by it; also the instruction of leadership qualification; and the measures for public ownership, all began a new series of deliberately socialist policy imitatives. It was accepted by the TANU (Tanzanian African National Union) Party, headed by President Nyerere in 1967, *Ibid.*, p. 231.
6. Mahatma Gandhi (1939), *Collected Works*, Vol., 68 Ministry of Information and Broadcasting, New Delhi, 1977; p. 266.
7. Mao Tse-Tung (1949), *Selected Works*, Vol. IV, Foreign Languages Press, Peking, 1961; pp. 453, 459 (22n).
8. F. Harbison and Cy. Myers, *Education, Manpower and Economic Growth* (New York: Mc Graw Hill, 1964), p. 185.
9. Don Adams, *Education and Modernisation in Asia* (New York: Addision Wesley, 1970); p. 109.

   For detailed analysis, *see* Sures Chandra Jain, *Education for Socio-Economic Development*, Concept Publishing Co., New Delhi, 2004.
10. Mahatma Gandhi, *Collected Works*, Vol. 65, Ministry of I&B, New Delhi, 1976; p. 129.
11. Rene Dumont, "A World Gone Mad", in CERES FAO of the U.N. Italy, Nov.-Dec., 1973; p. 19.
12. For Education for Self-Reliance, see Julius K. Nyerere, *Freedom and Socialism* (Dar es Salaam: Oxford Uni. Press, 1968); pp. 267-90.
13. Harbison and Myers, *op. cit.*, p. 62.
14. See Rowl Alley, "Building a Socialist Educational System in China:" Far Eastern Reporter, New York, 1974; p. 3.
15. See Elements of Educational Planning, Karus reprint, Nendch, in arrangement with UNESCO, Educational Studies Document No. XLV., 1971; p. 12.
16. John Hanson and Cole Bremback (Eds.), *Education and the Development of Nations* (New York: Hold, Rinehart and Winstan Pub., 1966); p. 120 editors comment.
17. For detailed analysis, *see* Sures Chandra Jain, *Education for Socio-Economic Development*, Concept Publishing Co., New Delhi, 2004.
18. Professor Gunnar Myrdal's Asian Drama: *An Inquiry into the Poverty of Nations*, is a massive three-volume study about poverty in South and South-East Asia, Published by Pantheon (New York) in 1968.

    Adam Smith's *Wealth of Nations* was published in 1776 in England.
19. John Hanson and Cole Bremback, *op. cit.*, p. 118.

20. V.K.R.V. Rao, *Education and Human Resources Development* (New Delhi: Allied Pub., 1966); p. 33.
21. A Freeman Butts, "*Teacher Education: A Focal Point*", in John Hanson and Cole Bremback, *op. cit.*, p. 374.
22. Chapter 3-21. The translation is from Swamy Chimayanand, *The Holy Gita.*
23. Educational Planning, UNESCO, Paris, 1970, p. 194.
24. V.K.R.V. Rao, *op. cit.*, p. 146.
25. Gurukul is a Sanskrit word; "Guru" means teacher "Kul" means residence or living place or Ashram.

    In ancient India, in the Vedic period, free education was well- known and was available to all students' regardless of their financial status without any payment by rendering personal services to the Guru.
26. G. Myrdal; Asian Drama: *An Inequiry into the Poverty of Nations*, Vol. III, New York: Pantheon, 1968); p. 1736.
27. V.K.R.V. Rao, *op. cit.*, p. 17.
28. See Chapter 2 in this book, "*Agricultural Mechanisation and Rural Industrialisation.*"
29. For detailed analysis, *see* Sures Chandra Jain, *Education for Socio-Economic Development*, Concept Publishing Co., New Delhi, 2004, especially the Chapter on "Inferiority Complex: the servile invitation of the Colonial Rulers of the West".
30. Editorial in *The Hindustan Times*, New Delhi, 28.11.1989; p. 10.
31. Elements of Educational Planning, *op. cit.*
32. Dam Adams, *op. cit.*, p. 154.
33. For details see F. Harbison and C. Myers, *op. cit.*
34. For detailed analysis, *see* Sures Chandra Jain, *Education for Socio-Economic Development*, Concept Publishing Co., New Delhi, 2004.
35. *Ibid.*
36. F. Harbison and C. Myers, *op. cit.*, p. 182.
37. See Chapter 2 in this book, "The formation of An Agro-Industrial Complex."
38. *The Hindustan Times*, New Delhi; 13.4.1988; p. 11.
39. In the process of rral development, the main theme of this book, the problem of abundant population, its geometric growth is serious problem. While attempts are being made to control the growth of the population, they can be realised only in the long run. The other immediate problems in the process of rural development, therefore, are manifold if we view the gravity of the situation in the immediate run.

    These remedial measures are discussed in various Chapters in this book.
40. *The Hindustan Times*, New Delhi, 20-7-1989; p. 11.

41. Pierre Pradeavand, "*The Malthusian Man*", in New Internationalist Devon Press, Berkshire, England, May 1974; p. 13.
42. Adopted at the Second Asian Population Conference, Tokyo, on November 13, 1972.
43. On September 20, 1972, the then Secretary-General of the United Nations, Dr. Kurt Waldheim, issued a statement formally proclaiming the year 1974 as "World Population Year". *Source:* Centre for Economic and Social Information, United Nations, New York; Pamphlet No. CES/WXY 8...73-03257.
44. *Ibid.*, para 1.
45. *Economic Times*, New Delhi; 3.12.1989; p. 11.
46. Alex Marhel, "The Unwanted Child of the U.N.", in New Internationalist, Devon Press, Berkshire, England, May-1974; p. 32.
47. Pierre Pradervand, "The best pill is development", CERES FAO, Rome, Italy, Nov.-Dec.- 1973; p. 8.
48. Marshal Wolfe, "The Day of Laissez-Faire is Done", in CERES, *Ibid.*, p. 45.
49. *The Hindustan Times*, New Delhi; 13.4.1988; p. 11.
50. Wm. S.W. Lim, "The Anticity", in CERES FAO of the United Nations, Rome, Italy, Nov.-Dec.- 1973; p. 61.
51. *Ibid.*, p. 62.
52. See Chapter 2, "Formation of an Agro-Industrial Complex for Agricultural Mechanisation and Rural Industrialisation", and Chapter 5," Utilisation of Natural Resources."
53. Paul R. Ehrilich, *The Population Bomb* (New York: Ballantyne Books, 1971); front page cover citation.
54. Rene Dumont, "A World Gone Mad," in CERES magazine, FAO of the UN, Rome, Nov.-Dec., 1973, p. 17.
55. Quoted by Authony R. Messham, "Neither Marx nor Malthus", CERES FAO of the United Nations, Rome, Italy, Nov.- Dec.- 1973; p. 40.
56. Rene Dumont, "*A World Gone Mad*", in *Ibid.*, p. 17.
57. Robert L. Heilbroner, *The World by Philosophers* (New York: Siman and Schuster Pub., 1972); p. 76, Especially See Chapter 4, "Gloomy Presentiments of Parson Malthus and David Ricerdo."
58. Angelo Abgelopoulos, *op. cit.*, p. 58.
59. Lester B. Pearson (Chairman of the Commission), *Partners in Development* (New York: Praeger Pub., 1969); p. 58, para 6.

    "This Commission on International Development" was authorised by the World Bank (I.B.R.D.) to prepare a report on the poverty of nations in the Third World and its relations with the Western countries. Pearson was the Prime Minister of Canada, a Nobel Laureate. In 1980, came another such report, headed by the former West German Chancellor, Willy Brondt, North South: A Programme for Survival (Cambridge, Mass.: The MIT Press, 1980). Brandt was the head of the Independent Commission for International Development.

60. See Chapter 2 "Formation of an Agro-Industrial Complex: Agricultural Mechanisation and Rural Industrialisation."
61. Peter Adamson, "An Introduction to the World Population Year, 1974," in New Internationalist, Devon Press, Barkshire, England, May, 1974; p. 8.
62. Z.P.G. or Zero Population Growth is a non-profit politically activated organisation dedicated to bring about population stability. Its national offices are in Washington, D.C., and Los Angeles, California. See Paul Ehrlich, Population Bomb, (New York: Ballentime Books, 1972).
63. Declaration of population strategy for development adopted at the second Asian Population Conference, Tokyo, Japan, on Nov. 13, 1972. See a United Nations Pamphlet, "1974 As World Population Year", New York, document No. CESI/WPY-8-73-03257, para 1.
64. For detailed analysis, *see* Sures Chandra Jain, *Globalization, Debt, Arms and Reforms in South Asia*, Concept Publishing Co., New Delhi, 2004.

5

Utilisation of Natural Resources

"A poor man does not use money as a weapon. Major means of production are: the land, forests; mineral resources; water; oil and industry, upon which a large section of population depends for their living, or which produces essential raw materials."

—Julius K. Nyerere[1], 1967

...Poor people are active agents in their lives, but are often powerless to influence the social and economic factors that determine their well-being. ... The world has deep poverty amid plenty. Of the world's 6 billion people (in the year 2000) 2.8 billion-almost half -live on less than $2 a day, and 1.2 billion-a fifth-live on less than $1 a day, with 44 per cent living in South Asia.

—World Development Report, 2000/2001[2]

A. INDIGENOUS NATURAL RESOURCES FOR ECONOMIC DEVELOPMENT

This chapter largely deals with the planning of development projects such that the necessary variable inputs are at the disposal of the individual nation. In case of India, these variable inputs are people, land, water, forest, minerals, energy resources, etc. In labour-intensive projects, how to utilise these variable resources are discussed in greater detail. The World Bank in its, World Development Report 2000/2001 mentioned that:[3]

> ...Poor people are active agents in their lives, but are often powerless to influence the social and economic factors that determine their well-being. ... The world has deep poverty amid plenty. Of the world's 6 billion people (in the year 2000) 2.8 billion—almost half—live on less than $2 a day, and 1.2 billion—a fifth—live on less than $1 a day, with 44 per cent living in South Asia.

The main emphasis in this chapter is AN outlining certain labour-intensive agro-based small-scale industries, where the bulk of Indian unemployed masses, particularly in rural areas, could be utilised. V.P. Singh[4], when he was Prime Minister, pledged to make the "right to work" a fundamental right to be guaranteed under the Constitution.

It is obvious that today's geometrically expanding population, if not curbed, can bring social and cultural disaster and the degradation of humanity, not very long from now. In addition to the human-made and human-controlled social phenomena of population expansion in the under-developed nations, unrestricted consumption of natural resources can bring disaster even earlier. On the other hand, if the natural resources are not developed and extracted, it will worsen the problem even more. Natural resources are certainly not infinite. A faster industrial growth means more consumption of the natural resources, which, in turn, means that the natural resources are also being exhausted at a faster rate. But, unfortunately, the industrially developed nations, which are consuming natural resources at a faster rate to feed their industries, have been doing so at the expense of cheap natural resources of the less industrially developed poor nations. During the colonial period of history, there were no 'question of a fair commercial return of the colonised nations' natural resources. After the disintegration of the colonial system, the commercial return for the raw material of the under-developed nations has been less than it should be, as the under-developed paid more for the manufactured goods of the developed nations.

All this is quite evident from the Sixth Special Session of the General Assembly of the U.N., New York (held on 9 April-2 May 1974), on the problems of the raw materials of the under-developed nations. I had a chance to observe some of these meetings and I found that almost every delegation from the under-developed and the developing nations complained about either not being paid a fair price for their natural wealth or the indirect control of their economies. Some delegations even said the issue was a threat to the international peace and security.

Mr. Paul[5] Rao proposed that :

> "The General Assembly should declare that actions perpetrated by States, multinational enterprises, or any other entity against a State be determined to make deep structural changes in the economic and social realms or to exercise permanent sovereignty over its natural resources, were contrary to the principles of the UN Charter and constitutes a threat to international peace and security."

This statement means that a mechanism and structure for the international economic system is now needed to stabilise the world money market. Mr. Swarn Singh[6] of India said that the present economic system of the world is inherited from the past, in which colonialism and the other forms of exploitation were the dominant features. Mr. Shevel[7] said the same thing the other way round: The activities of the multinational monopolies were the principal obstacles to de-colonisation and constituted the undisguised robbery of people suffering under the yoke of colonial economy.

The unfair price relationship between natural resources imported by the developed nations and machinery and other sophisticated technical equipments imported by the under-developed nations is, in some way, exploitation. Table 5.1 shows an example of the unequal exchange:

Table 5.1: Unequal Exchange between Developed and Developing Countries

To buy one tonne of Steel	*1951*	*1961*	*Increase*
Malaysia (1bs. rubber)	132	441	334 per cent
Brazil (1bs. Coffee)	158	380	240 per cent
China (1bs. Cocoa)	202	571	283 per cent

Source: Jack Woddis, *Introduction to Neocolonialism* (New York: International Publishers, 1971); pp. 90, 94.

In the past quarter of the century[8], a large American rubber firm, has taken $ 160 million worth of rubber out of Liberia, in return the Liberian Government has received the paltry sum of $ 8 million. The average net profit made by this American company is three times the entire Liberian revenue. What do the above examples indicate?

Discovery, development, and extraction of natural resources are essential for economic development. Under-utilisation of the natural resources has a direct effect on all national economies as far as development is concerned. The process of discovery, development, and extraction of the natural resources, no doubt, is a complicated technical manipulation and often requires sophisticated technologies. However, it does not mean that these nations cannot make full utilisation of their natural resources. The multinational corporations have been doing this job in the under-developed nations. Unfortunately enough, the interests of the host under-developed nation have usually been ignored, and she has often become the parasite of the investing multinational corporation. This is clearly visible from the Table 5.1 Professor Moorse suggested some constructive ideas and practical approaches to reduce the hazards and frustrations on the part of both, the investors and the poor[9] host nations:

1. Proposals for ameliorating or offsetting the effects of instability in primary commodity rate;
2. The trend towards establishment of international joint business ventures;
3. The use of international arbitration procedures;

4. Guarantees for international investors; and
5. Possibilities of standardizing legislation relating to minerals and other natural resources.

Of these suggestions, as far as the establishment of international joint business ventures, international arbitration procedures, and the legislation relating to the minerals and other natural resources are concerned, they already exist in one form or the other either under the auspices of various UN specialised agencies or through bilateral agreements. Conflict of interests arises only in the guarantees for international investors and the effects of instability in primary commodity rate of the poor host nations. Whenever steps are taken to nationalise any industry in any poor nation, it is largely because of the unequal price[10] relationship between the raw materials and the manufactured goods, as explained in Table 5.1. Thus it can be said that the guarantee of the international investors lies largely in themselves, if they do not practice this unequal and unfair price relationship.

The discovery, development, and extraction of the natural resources, however, does not in itself guarantee the economic growth of the poor nations. The decisive factor is its usefulness within the country itself or the potential world market. It can be interpreted as an interaction between the economics and politics of the nations involved.

There are various stages involved between discovery and extraction. The inputs in these stages are usually heavy capital investments and a greater part of skilled and technical personnel. This heavy capital requirement invested by multinational corporations has not been able to contribute very much to the poor host nation, because,[11] "remittances of profits and amortisation charges and the procurement of equipment and supplies abroad have left the host country with only part of the export proceeds." The recent Sixth Special Session of the General Assembly of the UN, as mentioned earlier, in its resolutions has not yet create an environment to provide stability in the world market and fair prices for the raw materials anyway.

Virtually everything has earlier been said either in any of the UNCTAD I, II, or III, or was formulated in "The UN International Development Strategy" for the seventies.

Similar recommendations were made by some delegations to the UN Committee on Natural Resources, 1973. One of them was[12],

> "The concept of permanent sovereignty over the NRs should be applied in such a way that the countries with NRs and countries with capital and technology could cooperate to maintain mutually complementary links. The Committee also emphasised that the needs for profits derived from investments to be reinvested in the host country in order to promote its development, allowing however for the repatriation of reasonable profits in conformity with the principle and mutual benefit."

Consideration must also be given to the establishment of a World Resource Development Institute[13]. "The direction of such an Institute could be by an international board of world leaders in the resources field, of scientists, engineers, planners, economists, administrators, and educators". This cooperation can succeed when the needs and the resources of each and every nation are considered according to their requirements, because: (1) the natural resources are very unevenly located; and (2) not every nation that possess them has enough technical capability and know-how and skilled manpower for their development and subsequent extraction.

Here, it will be pertinent to recall that it is the industrially developed nations of today who paid a heavy toll in terms of human misery and human lives during the earlier phases of industrialisation, especially in building the rail-roads and in exploration and adventures in mining. The underdeveloped nations of today were, at that time, under the colonial rule, and were exploited almost as unpaid workers on volunteers. In this sense, both the developed and the underdeveloped nations together deserve their respective share: the former for their efforts, and the latter for not being given a chance to participate in the process of industrial development. Having both these factors in mind[14] "they (underdeveloped) can avoid the mistakes made in Europe and North America, they can take short cuts".

The natural resources contribute a great deal to the national economic growth. They: (1) either provide raw material to the nation's industries or (2) earn the scarce foreign exchange and tax revenues in case natural resources are sold in world market. The review and appraisal of the NRs falls into two categories; one, human want and second, the human art to overcome the want. The former (human want) is related to consumption for a better living. The latter (human art) deals with discovery, development, and extraction of the NRs and the necessary inputs that it requires, *i.e.*, capital investment which is usually a heavy initial outlay, and the human resources which are largely skilled and trained. Thus the two sets of criteria may be distinguished as follows[15]:

1. The state of the wants, the expression of human needs, individual wants, social objectives, higher education; and
2. The state of the arts, as the summation of the human capabilities along lines of both technology and social organisation.

Thus the appraisal of the NRs is a joint responsibility and it proceeds from:

1. The knowledge of facts of nature and culture to,
2. The determination of technical feasibility to,
3. The formulation of a grand strategy along the socio-economic lines.

I would, however, prefer self-reliance as the objective, categorise the NRs only in the fields which the underdeveloped nations can develop on their own and which do not need heavy initial capital outlay and sophisticated technology.

B. LABOUR INTENSIVE PROJECTS INVOLVING FORESTRY

Of the available area of earth, about one-thirds (roughly 4400 million hectare) is covered with one or the other types of forest. But only one-thirds of this available area is in use and more than another third, at present, is considered inaccessible. Nearly one-half of the world's forests are concentrated in the USSR and Latin America alone, where the extensive areas of virgin forest still exists. In the under-developed nations, there is growing pressure over acquiring the forestland for agricultural purposes. There are not effective and adequate regulations for preservation of land for forestry. Those who hold the rights of ownership of the forestlands, try to clear the forestland for agricultural and industrial purpose.

Table 5.2(a): Area under Forests (lakh hectare)

Description	*1973-74*	*1974-75*
1. From point of view of Exploitation:		
(a) Exploitable (forests in use)	428	420
(b) Potentially exploitable	178	181
(c) Others	144	148
Total	750	749
2. By ownership:		
(a) State	713	717
(b) Corporate bodies	24	21
(c) Private individuals	13	11
Total	750	749
3. By legal status:		
(a) Reserved	381	391
(b) Protected	242	239
(c) Un-classed	127	119
Total	750	749
4. By composition:		
(a) Coniferous	42	47
(b) Non-coniferous (including bamboo)	708	702
Total	750	749

The forest resources of India, in 1973-74[16], amounted to approximately 75 million hectare. Table 5.2 shows the area under forests with respect to exploitation legal status, and by composition. Forests occupy about 22.8 per cent of the total geographical area. The annual recorded production of industrial wood and fine wood, including wood for charcoal, separately for coniferous and non-coniferous species during 1973-75 is indicated in Table 5.2. The official policy in India is to maintain approximately one-thirds of the total geographical area under forests. But pressure continues to build on the forestland, due to cultivation and projects involving river-valley developments. The present policy in the country is to stop deforestation, and measures are taken to preserve the forestland. The object of forestry development[17]:

Table 5.2(b): Production of Industrial and Fuel Wood

Description	*1973-74*		*1974-75*	
	Quantity (lakh receipt)	*Royalty*	*Quantity (lakh receipt)*	*Royalty*
	Cubic Metres	*(Rs. Crores)*	*Cubic Metres*	*(Rs. Crores)*
1. Industrial wood:				
(i) Coniferous	10		10	
(ii) Non-coniferous	87		88	
Total	97		98	
2. Fuel wood (including wood for charcoal)		132.60		178.63
(i) Coniferous	4		4	
(ii) Non-coniferous	161		161	
Total	165		165	

Actual market value of the major forest products is consderably higher than the royalty received by the State Governments.

Source: India: A Reference Manual, Publications Division, Government of India, New Delhi, 1977-78; pp. 223, 224.

"Has been the maximum supply of various forest products on sustained yield basis, particularly of industrial and commercial timber and fuel wood so as to meet the increasing requirements of the country. In the Fifth Plan, special emphasis had been laid on

(i) nationalisation of forestry operations,
(ii) protection and management of wild life,
(iii) commercialisation of industrial forestry, and
(iv) environmental and social forestry.

With this aim in view, a dynamic programme of production forestry, by creating large-scale man-made forests with the help of institutional financing, has been launched to achieve a substantial rise in the production of industrial wood. Measures to augment fuel wood resources and small wood for farmers through social forestry schemes have also been initiated.

The elimination of contractors in some States has greatly reduced illicit felling of valuable trees. Side by side, new planting programmes are taken up. Other important central schemes in operation include forest research, pre-investment survey of forest resources, and logging training centre project."

It requires hundreds of years to develop a forest soil when it has been destroyed, either by clearing or by fire. The demand for forest products is increasing steadily, particularly for use as industrial raw materials. In a Preface to a study, "World Trends and Prospects: Wood", the Director-General of the FAO[18], said

"The evidence assembled in the study points at the rapid rate at which the industrial wood sector is growing. In the decade of 1961, the quantity of wood used annually for industrial purposes grew by about 25 per cent; in subsequent 14 years, it will probably grow by nearly 45 per cent. Demand is shifting from unprocessed to processed wood products, and, among the latter, from solid to reconstituted wood products such as particle board and paper."

This is the kind of industry, which the under-developed nations can develop without any considerable external assistance. At the same time, there is a favourable international market for forestry-oriented industrial goods.

The poor nations should focus their attention on forest industrialisation. It will mean that the same area of new plantation must replace the area of the forest that is being cut for industrial purposes. Improving logging and transportation facilities is one of the most important factors in making forests' cultivation economical. A heavy initial outlay is needed in the beginning to provide the necessary infrastructure of paved roads and railways for transportation. This, however, can be built gradually, although it might take a little longer. Accessibility to the forest areas is a primary function.

To meet the challenge of growing population, to provide enough information, and to stop further crowding of the already thickly populated cosmopolitan centres, it is necessary to provide and maintain the logging and transportation facilities and to exploit the inaccessible areas of forest land. A United Nations study[19] shows that a yield of only one cubic metre per hectare per annum from only half of the 2,900 million hectare at present unused would result in an enormous increase in production— about double the present world production. The initial yield in many virgin forests, especially in some large areas of tropical Africa and Asia, might be much larger.

In Latin America for a few years, almost 20,000 kilometre a year of permanent roads have been built.

There is a great need in America for forestry and plans are being worked out regarding logging and transportation facilities. The forest industry is a remote return as far as economy is concerned. Spending money today on research, scientific study of composition and structure of forestry, on soil mechanics of the area to be planted, is in the national interests, though only in the long run.

As pointed out in Chapter 4 (Mobilisation of Human Resources), the underdeveloped nations are on the brink of facing catastrophic population growth with the present birth rate, and thus the job-creating and labour-intensive forest industry, both for clearing and planting of the new forests, could be an additional advantage. Though much of the forestland in the Western States (the U.S., Europe, and Japan) is privately owned, in the under-developed regions of Asia, Africa, and Latin America it will be a better solution to formulate a policy to bring forest control under government rule. This will help facilitate and finance the initial outlay, which otherwise may not be possible. This is so, because the privately owned forestland is in small holdings, which make it uneconomical to merchandise.

The development of forest industry provides, in addition to what has been said so far, the following industries:

1. Pulp and paper industry,
2. Saw mill industry,
3. Fibre raw material,
4. Wood-based panel industry.

Having realised the importance of forest industry and the extent of the forest resources of the world that have not been utilised and lie in the underdeveloped nations, the Seventh World Forestry Congress, which met at Buenos Aires[20] in October 1972, recognised the contemporary challenge in this direction to expand this industry and recognised the need for the expansion of share of developing countries in international forest products trade.

Water: Surface and Underground

This natural resource has for thousands of centuries been a catastrophic and devastating factor in the economies of nations around the globe. It has created havoc in uncontrolled and non-planning States. This is particularly true of South East Asia and certain regions in Africa; either in the from of (1) Droughts, or (2) Floods. To utilise this resource, the above two factors have to be dealt with by: (1) Irrigation networks to combat droughts, and (2) Construction of dikes and dams to curb floods. The fresh waters comprise less than one per cent of the earth's total water supply; the rest is salty ocean waters or tied up in glaciers and the polar ice caps. Of this limited available fresh water supply a significant portion is being wasted in floods. This water could well be used to combat droughts through water diversion planning and to provide irrigational facilities, the very basic requirement of agriculture. In addition to this wastage in floods, about 100 cubic kilometres of water, which daily runs off into the oceans, needs a careful consideration for its effective use for the benefit of mankind.

Underground water constitutes a major resource. Until recently, with no mechanical power being discovered to drill wells at greater depth, it has largely been dependent on muscle power. Throughout the wide belt of arid lands, droughts and scarcities are a perennial problem. Water resource development is a sustained yield of available water to meet the demands for increased fresh water as the under-developed nations are progressing towards agriculture and industrialisation. Recently, India, with financial aid and technical assistance from the UNDP[21] (United Nations Development Programme), has undertaken a project to use the excess waters of the rivers of Ganges and Brhamputra to irrigate the vast central region by creating a 3,000-kilometre canal. This project, when implemented, will be the world's largest water diversion schemes. About 25 billion cubic metres of water would annually be pumped from the rivers to the reservoirs.

Water resources development planning takes into consideration many factors, the most important being the geographical location. With planning, all possible sources of availability and supply (the annual rainfall, and the quantity of water available on surface and underground) can be assessed. The next factor is the amount flowing out as waste and the amount affected by pollution. After these factors comes the question of amounts to be allocated. The allocation must be planned, largely on a long-term basis, ranging from 10 to 20 years, or even more. The allocation needs a careful coordination of the country's requirements, namely, industrial requirements, domestic supply, power production, and agricultural needs. Though practically it is not possible to get good and reliable data for estimation for allocation purposes, due to the fact that evaporation and transpiration consume significant amounts of water in hot and dry areas even before utilisation by the crops, a rough estimate would suffice.

To provide fresh water by desalinisation is a costly process, which could involve heavy initial outlay with slow commercial return over a long period of time. The recent advancements and developments in technology have, however, decreased the cost of desalination substantially, whereas formerly desalination used to cost about $ 1.06 per cubic metre. About 20 years later, it costs now roughly $ 0.33. But much more has to be accomplished before fresh water, converted from the saline sources, will be cheap enough for irrigation and industrial purposes. The major factor is the cost of electric energy, which, unless reduced, will mean that some other new methods have to be developed. The cost of freezing, distilling, and electro analysis processes have been reduced substantially, as a result of continued technological innovations and research. However, the availability of fresh water from desalinisation can only contribute to the nations, which have coastal regions; and it is the ground water that offers greater promise both practically and economically. This applies particularly to the developing countries[22]—

> "Many of which are situated in arid or semi-arid zones where rain is rare and irregular, where the short and violent run-off washes down solid matter and is often full of dissolved salt and where intensive evaporation removes a large proportion of the rainfall and substantially reduces possible reserve capacity,...ground water has many advantages; it is constantly available, it is little affected by weather, it is protected from evaporation and it is usually of better quality, chemically and bacteriological, than surface water."

The river basin development in coastal regions needs a considerable amount of scientific knowledge and regional environmental studies, in addition to initial heavy outlay. A recent UN study discusses this problem in greater depth, with regard to public investment in water management and UN involvement in promoting both national and international action in river basin development. The study suggested that steps at international basis may be grouped under five classes[23]:

1. Improving basic services in hydrology;
2. Sharpening the tools for analysis and for concrete action in water resource utilisation;
3. Stimulation of scientific and technical investigation;
4. Aiding countries in their river basins; and
5. Laying the groundwork for reconciliation of conflicting interests over river basins of an international nature.

The study further suggested that the loans, etc. may be obtained from the IBRD (World Bank) to execute the programmes. The implementations of such international river basin plans on the assumption that they can be made possible through international assistance is similar to the notion of "*Let them eat cake.*" A representative in the UN Committee on Natural Resources, pointed out that[24], "*the outcome was a political document based upon mathematical calculations and that the result was a naive*[25] *attempt to analyse political issues by simple mathematical processes.*"

Minerals

Minerals are exhaustible, irreplaceable, and certainly not infinite. A study[26] shows that at the present rate of consumption, the consumption of metals in the next 35 years is expected to exceed that of the last 2,000 years, and energy use in the next 20 years is estimated to be three times that of last 100 years. This consumption is very uneven among the nations. The U.S. alone, with a population of merely 6 per cent of the world, consumes about 37 per cent of energy. The earth's available energy and minerals cannot support the entire population, which is estimated to be 3,700 million in 1974, even at the European rate of consumption. The underdeveloped nations are now constantly struggling hard and are accelerating their efforts towards industrialisation. This will mean that the consumption of NRs would be at a faster rate, because the industry feeds upon NRs and energy. The harsh realities of this phenomenon can neither be underestimated nor ignored. What will happen is exactly what has been said that[27]:

> "Unfortunately mineral resources are exhaustible, from which it follows that the faster a nation grows in industrial strength and military potency—a growth made possible largely through increased mineral output—the faster it liquidates the very basis of its power."

From the discussion so far, I may propose the following three alternatives:

1. Ignore the desire and the necessary requirements of the underdeveloped nations, as it has been up until now;
2. Be humanitarian and the under-developed nations are provided the share of NRs on an even basis and exhaust the NRs reserves;
3. Accelerate the efforts for discovery, development, and extraction of existing means of NRs and search for non-conventional sources of energy.

The first alternative can no longer be applied. The peoples of Africa, Asia and of Latin America will no longer allow themselves to be oppressed and their desires suppressed. The colonial time is gone, never to return. Hence the consequences of this alternative pose a serious threat to international peace and security, if the present trend is continued and these peoples are ignored. Alternative number 2 speaks for itself. This is not a practicable idea if we do care about the coming generations. They will certainly not 'forgive' us for our being so 'selfish'. A person from an industrialised nation, consuming the NRs at the above mentioned rate, may well 'smile' while passing by the rural areas of, say, my country, India, or any other underdeveloped nation, when he happens to watch the people using cow dung for heating and cooking instead of an electric furnace, using coconut shell instead of a metal pot, etc. Yet his children's children may have to resort to these resources, because of exhaustion and depletion of these NRs, for he is consuming these natural resources at a high rate.

This means that only alternative number 3 remains. The third alternative means that under-developed nations have to accelerate their efforts for discovery, development, and extraction, *i.e.*, to set up mining. At the same time, they have to find new ways to recycle the used metal and have to be economical in consumption. These raw materials constitute the foundation of almost every industry: construction, electric energy and its distribution, communications, and agriculture. In sequential form, the fossil fuels create energy which runs industry, provides the consumer goods, a better and comfortable living. To achieve this goal, we need a new breakthrough in technology, deploying scarce and limited means available at our (underdeveloped nations) disposal. I may categorise this process into two main areas for an effective exploitation of the NRs:

1. National level organisation;
2. Survey and discovery (including cost/profit analysis).

In dealing with NRs, the policy planning should be formed in realistic terms than in idealistic theories. This will mean that, if the prospective underdeveloped nation does not itself possess enough capabilities, a decision is needed as to whether the foreign firms should be allowed to extract the NRs. The feasibility of such a decision has to be realised only on a realistic basis. If the foreign capital investment is decided upon, there should be enough legislation to deal with the exercise of rights over the sovereignty of natural resources. If the foreign capital investment and technical know-how is undesirable, the government has to make a careful review of all the factors affecting the decision and its possible consequences in the long run. The foreign investors largely invest their capital to make profits, and when the repatriated profits begin to exceed the invested capital, then are not in national interests[28]. On the other hand, if foreign investment may seem, for a while, as the only available alternative, an optimal decision, considering the opportunity costs, must be made.

The second criterion, on the national level organisation for an effective exploitation of the NRs, is that geology and mining, as far as possible, must be under one roof with closer coordination and effective bureaucratic functioning. The principal function of the geological institutions must be to make an assessment of a nations' metallic and non-metallic material, and of fossil fuel, for annual statistical purposes to help facilitate mining operations.

In many under-developed nations, adequate well-trained personnel may not be available. The main drawback about the training of the mining personnel is that this training cannot be provided on the job. In many under-developed nations, there is not only a shortage of training geologists and mining personnel but also of the facilities for their training. This is a complicated problem in two ways: (1) if there is only a small amount of NRs available within the country, the demand should largely be filled by either employing specialists from aerial survey companies, or from the UN agencies or by any suitable bilateral agreements with any country; and (2) if the country has enough NRs expected to last for about a decade or more, then policies should be formulated from a different perspective. This could be done by getting the personnel trained in a friendly country. In geological and mining operations, one must bear in mind whether the personnel trained abroad would find similar conditions at home for which they may or may not have been trained.

One of the prime factors in the field of NRs is the detailed study of the nation's geology. The maintenance of National Geological Survey to collect all the pertinent information about geology is also significantly important in the selection of sites for dikes and dams (for hydro-electric power and irrigation purposes), quarries and gravel pits for road construction, and underground water for agricultural and domestic supply purposes. To help facilitate survey and scientific study of the nation's geology, and to make an optimal use of available resources, we need certain tools: mapping of rock units, photo geology, paleontology, petrology, geophysics and geochemistry. Photo geology is a sophisticated technology and the decision to choose either this or the conventional methods largely depends upon the nation's financial status and technological capabilities. The recent advancements in airborne geophysical surveys are an even more sophisticated scientific tool, but it has certain advantages over later discoveries, such as magnetic permeability, nuclear radiation, and electromagnetic induction. This consequently increases the probability of discovery of these materials, such as magnetite, uranium, or the supine ores. It is the evaluation, judgment and interpretation of the person in charge to carry out these operations to make a rough guess as to what extent there is any possibility of any mineral deposit. These operations could be planned either on a short-term or on a long-term basis, depending upon the probability of ore deposits.

To make a reliable rough estimate on low, medium or high investment, depending upon the probable discovery of ore, a very high quality scientific infrastructure is needed at a fairly early stage for analyzing and testing of the samples. Metallic materials require heavy investment both at an early stage and, if discovered, for extraction. Energy is needed to extract the minerals from ore, which imposes an extra constraint, especially on the developing nations because of the hikes in oil prices[29].

While investment is heavy in survey and discovery, and subsequent extraction of the metallic materials, the non-metallic materials such as potash and other salts, clays, glass sands, gypsum, building stone, etc., require relatively low investment at all stages (survey/and discovery, development, and extraction). At the same time, these non-metallic materials find an immediate use by local industries, and thus do not have much in transportation costs. For an underdeveloped nation, therefore, when planning for the industrial sector, the discovery and development of non-metallic material is of great importance both from economic considerations and industrial development.

NOTES

1. Julius K. Nyerere, *Freedom and Socialism* (Dar-es-Salaam: Oxford University Press, 1968); p. 235.
2. The World Bank, World Development Report 2000/2001, Oxford University Press, New York, 2000 (The poverty line is $1.08 a day at 1993 PPP), p. 23.
3. *Ibid.*
4. *Hindustan Times*, New Delhi; 2.1.1987; p. 9. The new Government announced its time bound Action Plan on 1.1.1990. The 'Right to Work' was one of the programmes on priority Agenda.
5. Paul Roa, Foreign Minister of Cuba. U.N. 6th Special Session, New York, April-May 1974. Press release GA/5006, 24th, April, 1974.
6. Swarn Singh, External Affairs Minister of India, *Ibid.*
7. Mr. Shevel, Foreign Minister of Ukraine; *Ibid.*, press release GA/4993, April 19, 1974.
8. Jack Woddis, *Introduction to Neo-Colonialism* (New York: International Publishers, 1971); pp. 90, 94.
9. C. Moorse, "Potential and, Hazards of Direct International Investment in Raw Materials in Marien Clawson (Ed.) *Natural Resources and International Development* (Baltimore: John Hopkins Uni. Press, 1964); p. 398.
10. For other areas of unequal relationships between developing and developed nations, see the author's other book Globalisation...
11. J. Grun Wald and P. Musgrore (Eds.), *Natural Resources in Latin American Development* (Baltimore: John Hopkins Uni. Press., 1970): pp. 26, 27.
12. Committee on Natural Resources, Report on the Third Session, 6-17 February, 1973. (New York: United Nations Economic and Social Council official records, 54 session, supplement 41, 1973); pp. 19-20.
13. Science and Technology, Natural Resources Vol. II (New York: United Nations, 1963); p. 200.
14. A. Moyes and T. Hayter, World III (Oxford, England; Oregon Press, 1965); p. 33.
15. W. Zimmerman, *Introduction to World Resources* (New York: Harper and Row, 1964); p. 24.
16. These data are condensed from *India Year Book* (Reference manual) 1977-78, Publications Division, Government of India, New Delhi, p. 223 ff.
17. *Ibid.*, p. 224.
18. Quoted in G. Jones, *The Role of Science and Technology in Developing Countries* (New York: Oxford University Press, 1970); p. 94.
19. Freedom from Hunger Campaign, Basic study No. 10, "Possibilities

of Increasing World Food Production", FAO of the U.N. Rome, Italy, 1963, p. 213.

20. State of Food and Agriculture, FAO of the U.N., Rome, Italy, 1973, p. 49.
21. However, there are pitfalls associated with such types of projects being undertaken by International Agencies. See the author's other book, *Globalisation*.
22. Science and Technology for Development, Vol. II (Natural Resources), United Nations, New York, 1963; p. 63.
23. Integrated River Brasin Development, Report of a Panel of Experts, United Nations, New York, 1970; p. 40.
24. *Ibid*., Paragraph 134, pp. 28-29.
25. About these naive political attempts, see *see* Sures Chandra Jain, *Globalisation, Debt, Arms and Reforms in South Asia*, Concept Publishing Co., New Delhi, 2004.
26. Our Nation and the Sea, The United States government publication, Washington D.C., 1969. All the data are condensed from this document.
27. Natural Resources and Economic Growth: Resources for the Future. The United States government publication, Washington D.C. March 1961. These comments are from papers presented by Barnett, Potter and Christy, p. 80.
28. For detailed analysis, *see* Sures Chandra Jain, *Globalisation, Debt, Arms and Reforms in South Asia*, Concept Publishing Co., New Delhi, 2004.
29. Since the war in 1973 between the Israelis and the Arabs, the Arabs had increased the price of their oil. While it affected the rich Western industrial powers the many poor countries in the Third World simply could not get this so highly priced oil without severely hurting their already fragile economies.

6

Conclusion: The Removal of Poverty

A. Evolutionary process: The Policy of the British Colonial plunder
B. Reformist approach: Opportunism in the post-Independence era since 1947
C. Revolutionary change: The plight of the poor and exploited

"It is...evident that...capitalist development of the kind which helped countries of Western Europe in 19th century does not hold much promise for the countries of South Asia today. If that path is nevertheless adopted on account of the power of vested interests, as it has been in India today (despite all the socialist phraseology), the consequences can be nothing but disastrous."

—*K.N. Raj*[1], *1977*

...In all South Asian countries governance by personal rule has led to subjugation of the institutions of state to certain individuals and their coteries at different levels. As a result, the ability of the state to govern has grossly declined.

—*K.L. Kamal*[2], *1997*

The development and modernisation of underdeveloped societies (where from nearly 70 per cent to almost 95 per cent of their population live in the rural areas in hundreds of thousands of villages) clearly depends upon the social, economic and political awareness in the rural areas. The poverty-stricken population[3] in these villages, varying from a few hundred to a few thousand, must engage in productive work[4] and gainful employment. To this end, in the preceding chapters of this book, an attempt has been made to form an Agro-Industrial Complex with indigenous efforts based on land reforms, the mobilisation of human resources, and the utilisation of natural resources such as forest, waters, and minerals.

Since this chapter is primarily concerned with India, about 80 per cent of her over 844 million (1991 Census) people live in rural areas, in over half a million villages. In other words, the soul of India lives in rural areas, as Mahatma Gandhi used to say. Rural development is the development of the quality of the human factor. Nearly 90 per cent of India's rural population is illiterate, unable even to sign their papers, except by putting on the their thumb impression. The overall literacy rate is about 40 per cent. The improvement in the quality of human factor is a necessary prerequisite for other aspects of rural development to secure food and nutrition, housing, health, recreation, security, spiritual satisfaction, family planning, etc.

The problems and strategies for the uplift of rural areas for successful the implementation of projects are manifold. The rural development projects, as functions of planning and organisation, may be (a) short-term; (b) medium-range; and (c) the long-range objectives, depending upon the nature and function of specific projects involved. Besides, geographical and ecological factors, the institutional structure, there are a host of other factors that ultimately determine success on failure. The factors of innovation require the following, as stated by a United Nations document[5]:

> "To effect change, consideration must be given to the dynamics of change within the system; the extent to which the system can generate its own momentum for change, and the role of "external" agents in this change. Attention should be given to the training of rural development agents from within the community itself as well as those from the outside: welfare officers and assistants; doctors, nurses and medical and public and adult education officers, etc.

And further;

> Rural development programmes should be prepared democratically by consulting, and with the support of, the rural inhabitants who should participate in the selection of targets and in all phases of the implementation of the programme."

Rural development, as I see it, means the generation of employment by the utilisation of land, natural resources, and the mobilisation of rural population. The utilisation of the available rural work force for increased production and efficient productivity requires its education and training. In the case of India, the educational system[6] has not been able to re-vitalize the rural economy. The rate of return, both social and economic, is less in higher and urban education than in basic education in rural areas.

The Government of India created a new department of Rural Development in October 1974. The Department was made responsible for the work of the erstwhile Department of Community Development and also for work relating to agricultural credit and indebtedness, including cooperation in the agricultural sector. The main responsibilities of the Department include (a) all aspects of rural development including community development and Panchayat[7] Raj, (b) special programmes for the betterment of weaker sections of the community like the small and marginal farmers, the development of drought-prone and tribal and hilly areas, and rural manpower planning and employment, (c) agricultural credit and marketing including the development of regulated markets.

It should be borne in mind that planners, while making policy recommendations for rural development, must be aware of the fallacy of eclecticism that they can import policies and programmes that might have been devised and have been successful in foreign lands. The planning process for the development and utilisation of rural manpower ought to be consistent with the prevailing trends of the national needs and requirements and resources. But the planners in India, with westernised approach, cannot benefit from the experience of other nations such as Russia or China where the objectives and the policies are radically different. Professor Keith Griffin[8] says that though the:

> "Countries differ primarily in degree, not in kind; policies, ideologies and objectives are scattered along a spectrum in multidimensional space. Nonetheless, three distinct strategies or approaches to development in general and rural problems in particular can be detected: we shall label these the technocratic strategy, the reformist strategy, and the radical strategy. These differences in objectives, ideologies, institutions, and distribution constitute differences in style. Differences in style, in turn, are related to the classes or factions on whom the government depends for support."

Professor Griffin[9] summarised the various styles of rural development in various countries. Of these various styles of rural development strategies (technocratic, reformist and radical) the only countries where rural development projects have been successful are the ones who have had a radical rural development strategy, mainly China, Cuba, and Algeria. Mexico and Egypt have a reformist approach, and they have always met with miserable failure as we have been facing in India ever since Independence. It would be safe to suggest that the process of strategy formulation for generating employment in rural areas[10] must reflect the structure, institutions, and political philosophy of the country concerned. When technical and political constraints limit the choice in formulating the design and implementation of rural development projects, there are generally two schools of thought:

1. Centralised or an integrated management and control;
2. Decentralised or locally administered management and control.

When we consider the following factors in the mobilisation

of material and human resources in the rural areas, the integrated approach seems much more likely to yield results[11]:

1. Active participation of the population involves young people and women, in developing positive attitudes towards the improvement required, through community development, "animation rural", social service programmes, and other techniques;
2. Techno-economic characteristics including cost-return analysis and covering the various physical and economic factors relevant to rural development programmes, *e.g.* land, water supply for irrigation and other uses, energy, road and communications, crop production, and marketing facilities;
3. Human resources development needs comprehensive manpower surveys and assessment for vocational training and productive rural employment creation, orientation of education towards rural needs, promotion of health, nutrition, housing, and social welfare services;
4. Institutional development needs a sound institutional framework to support and sustain rural development, *e.g.*, land tenure and land ownership, appropriate public and semi-public services, voluntary rural associations and organisations, development of cooperatives and credit facilities; and
5. Administrative machinery, which will facilitate integrated approach for effective communication at all levels.

A well-integrated and coordinated administration, management, and control make much sense at the planning level in that the planning may be local, regional, or at the national level. But an integrated and coordinated management, administration, and control at implementation level seem to me rather highly impractical.

This chapter argues for India to develop and modernise her traditional society, because her intense drive towards social development, economic growth, and political awareness in the rural areas are not feasible without radically altering the countryside, and transforming it for the better. The process of rural transformation may be either reformist, radical, or revolutionary. The state of "emergency" declared in June 1975, and lasted for less than two years, was much of a reactionary approach, meant largely for the personal ends of the ruling elite. It was by no means intended for an authentic national socio-economic development of the country. India needs to improve the quality of human resources in rural areas. Without this factor, the most complex and sophisticated planning process though India may possess, her masses are doomed to a miserable life. The plans of the Planning Commission of India carries in them the seeds of their own destruction for two simple reasons:

1. The implementation of plans is far too much centralised in the hands of the Central Government. There is literally no meaningful and productive participation of the masses at the local levels in the implementation of developmental projects.
2. The masses in rural areas, being largely illiterate, are not able to keep themselves fully informed of the micro details of various developmental projects being implemented by government for their welfare.

This chapter based on analysis of poverty *versus* polity, defines the constraints and imperatives in the task of rural development. A systematic attempt is made to dialectically unfold the mechanism by which the rulers in India since Independence, and in most of the formerly colonized nations in the called "Third World," keep on promising reforms, never carrying out what they promise, and keep on perpetuating the *status quo, i.e.,* their corrupt and inefficient administrative management.

To initiate any desired socio-economic change there are generally three broad approaches as follows:

A. Evolutionary Strategy,
B. Reformist Strategy, and
C. Revolutionary Strategy.

These three approaches are graphically illustrated in Fig. 6.1 "Time Taken *versus* Social Change."

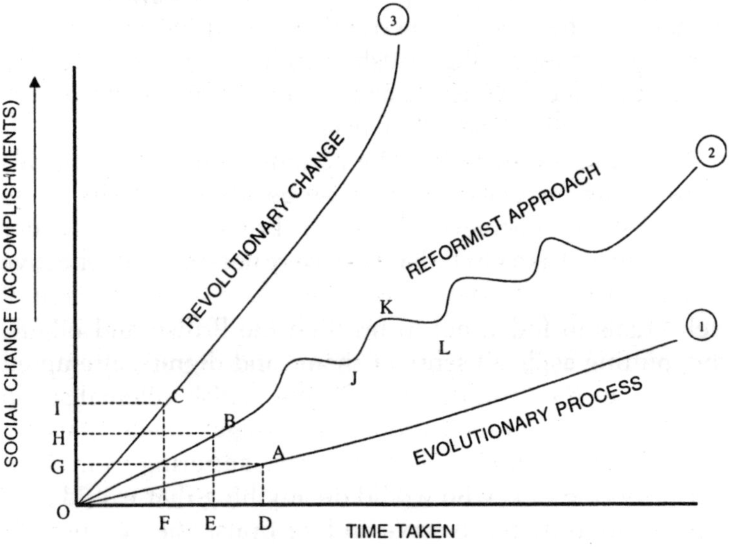

Fig. 6.1: Time Taken *versus* Social Change

This Graph illustrates TIME TAKEN *versus* SOCIAL CHANGE. Accomplishmensts based on the follwing three strategies:

Curve 1 The Evolutionary Process: The policy of the British-colonial plunder
Curve 2 The Reformist Approach: Opportunism in the post-Independence era since 1947
Curve 3 The Revolutionary Change: The plight of the poor and exploited

A. CURVE NO. 1: EVOLUTIONARY PROCESS: THE POLICY OF THE BRITISH COLONIAL PLUNDER

In the evolutionary process, as the term itself implies, it takes much more time (distance OD) to accomplish very little social change (distance OG) and some of the accomplishments in the evolutionary process, in fact, are rather by-products (the same distance OG) and are "necessary" for the effective plunder of the nation. A slave must not be starved to the point that he dies. This was the policy of the plunder of the British colonialism. The most educated and the most prosperous nation in those times before the British intruded, India was reduced to one of the most wretched nations during the heyday of the British imperialism and colonialism.

Nevertheless, there is some accomplishment during this evolutionary process (the distance OG). But the little accomplishments achieved during the British colonial plunder were not meant as a gift India. They were necessary for effective plunder, or else the slave could die. There is no shortage of intellectuals in India, not to mention the British and others, who, putting aside all sense of shame and dignity, attempted to point out the "contribution" of the British colonialism to the development of the nation. These vulgar intellectuals are nothing but schizophrenic flunkeys[12], suffering from deep inferiority complex, who would do anything that would give them an opportunity to serve and/or praise the "virtues" of their vanished colonial masters.

These intellectuals who attempt to assess the "contribution" of British colonialism to the process of development of the former colonies deliberately try to undermine the quintessential of such "contributions". In this regard, Karl Marx said as early as in 1853[13]:

> "The ruling classes of Great British have had, till now, but an accidental, transitory and exceptional interest in the progress of India. The aristocracy wanted to conquer it, the moneyocracy to plunder it, and the millocracy to undersell it. But now the tables are turned. The millocracy

> have discovered that the transformation of India into a reproductive country has become of vital importance to them and that, to that end; it is necessary, above all, to gift it with means of irrigation and of internal communication. They intend now drawing a net of railways over India. And they will do it. The results must be inappreciable."

Marx rightfully goes on to say that the British destroyed India by[14]:

> "Breaking up the native communities, by uprooting the native industry, and by levelling all that was great and elevated in the native society. The historic pages of their rule in India report hardly anything beyond that destruction. The work of regeneration hardly transpires through a heap of ruins."

The evolutionary approach, or the approach to eradicate poverty to which they were utterly indifferent, was the approach of the British colonial rulers. That is natural, tragic though it is from our point of view in India. The British policies were to plunder the nation, and this was their sole purpose.

Hence during the British period, no changes in rural areas were initiated, and if somewhere they were necessary to effectively plunder the countryside these changes were the undesired by-products. I do not intend to document this plunder by British imperialism any further, as I am primarily concerned in this book with an analysis of poverty *versus* policy in the post-Independence era.

After the end of colonialism in the aftermath of Second World War in 1945, the World Bank and the IMF stood as savior to provide loans and formulate programmes of socio-economic development in the developing countries, *i.e.* countries which were the colonies of the western countries before the Second World War. To maintain *status quo, i.e.* the poor countries continue to remain poor, and the hegemony of western powers remains unchallenged and unquestioned, in the former colonies, the World Bank and IMF prescribed "solutions" for the removal of poverty in developing countries. The World Development Report 2000-2001 of the World Bank has prescribed solutions as follows which are nothing but to perpetuate *status quo*:[15]

The World Bank strategy for poverty reduction

> The approach to reducing poverty has evolved over the past 50 years in response to deepening understanding of the complexity of development. In the 1950s and 1960s many viewed large investments in physical capital and infrastructure as the primary means of development.
>
> In the 1970s awareness grew that physical capital was not enough, and that at least as important were health and education. World Development Report 1980 articulated this understanding and argued that improvements in health and education were important not only in their own right but also to promote growth in the incomes of poor people.
>
> The 1980s saw another shift of emphasis following the debt crisis and global recession and the contrasting experience of East Asia and Latin America, South Asia, and Sub-Saharan Africa. Emphasis was placed on improving economic management and allowing greater play for market forces. World Development Report 1990: Poverty proposed a two-part strategy: promoting labour-intensive growth through economic openness and investment in infrastructure and providing basic services to poor people in health and education.
>
> In the 1990s governance and institutions moved towards centre stage—as did issues of vulnerability at the local and national levels. This report builds on the earlier strategies in the light of the cumulative evidence and experience of the past decade—and in the light of the changed global context. It proposes a strategy for attacking poverty in three ways: promoting opportunity, facilitating empowerment, and enhancing security.

Promoting opportunity

Poor people consistently emphasise the centrality of material opportunities. This means jobs, credit, roads, electricity, markets for their produce, and the schools, water, sanitation, and health services that underpin the health and skills essential for work. Overall economic growth is crucial for generating opportunity.

Facilitating empowerment

The choice and implementation of public actions that are responsive to the needs of poor people depend on the interaction of political, social, and other institutional processes. Access to market opportunities and to public sector services is often strongly influenced by state and social institutions, which must be responsive and accountable to poor people.

Enhancing security

Reducing vulnerability—to economic shocks, natural disasters, ill health, disability, and personal violence—is an intrinsic part of enhancing well-being and encourages investment in human capital and in higher-risk, higher-return activities. This requires effective national action to manage the risk of economy wide shocks and effective mechanisms to reduce the risks faced by poor people, including health- and weather-related risks.

The new poverty reduction strategy initiative of the World Bank and IMF[16]

The poverty reduction strategy initiative of the World Bank and International Monetary Fund seeks to link external support to domestically developed, results-based poverty strategies. It is also intended to improve the effectiveness of World Bank and IMF relations (and those of other donors as well) with recipient countries. As important as the recipient country strategy is the process leading up to it. A broad, participatory dialogue with representatives of civil society and the private sector is expected to:

- Help national authorities develop a better understanding of the obstacles to poverty reduction and growth and devise good indicators of progress in poverty reduction.
- Deepen a shared vision of desired poverty reduction goals across society.
- Lead to formulation of priorities for public actions to achieve the desired poverty reduction outcomes.

Encourage the development of participatory processes for setting poverty reduction goals and monitoring implementation and progress.

While the actual form of the strategy will be decided by the country—there is no single blueprint—most strategies would likely include:

- Long-term goals for key poverty reduction targets, and the macroeconomic, structural, and institutional framework for achieving them.
- Mechanisms for monitoring and evaluating progress towards the poverty reduction targets, linked to public actions.
- A consistent policy and institutional framework that includes the underpinnings for rapid, sustained growth and poverty reduction (including macroeconomic policies, institutional reforms, sector strategies, and associated domestic and external financing needs).

The World Bank considers, "four key areas of international action for poverty reduction:[17]

- Expanding market access in rich countries for developing countries' goods and services.
- Reducing the risk of economic crises.
- Encouraging the production of international public goods that benefit poor people.

- Ensuring a voice for poor countries and poor people in global forums."

Global and regional patterns: Income poverty and social indicators:[18]

- Between 1987 and 1998 the share of the population in developing and transition economies living on less than $1 a day fell from 28 per cent to 24 per cent. This decline is below the rate needed to meet the international development goal of reducing extreme income poverty by half by 2015.
- Because of population growth, the number of people in poverty hardly changed. But there are large regional variations in performance. East Asia and the Middle East and North Africa have reduced their numbers in poverty—East Asia dramatically so. But in all other regions the number of people living on less than $1 a day has raised. In South Asia, for example, the number of poor people rose over the decade, from 474 million to 522 million, even though the share of people in poverty fell from 45 per cent to 40 per cent. In Latin America and the Caribbean the number of poor people rose by about 20 per cent.
- Two regions fared particularly badly. In Europe and Central Asia the number of poverty soared from 1.1 million to 24 million. In Sub-Saharan Africa the number of poor people increased from an already high 217 million to 291 million over the same period, leaving almost half the residents of that continent poor. ... These variations in regional performance are leading to a shift in the geographical distribution of poverty. In 1998 South Asia and Sub-Saharan Africa accounted for around 70 per cent of the population living on less than $1 a day, up 10 percentage points from 1987.

Abhilesh Khare[19] wrote that:

"The IMF and the World Bank have focussed their attention

on India than any one else. With the assistance of the IMF and the World Bank, India is having river valley projects, environmental imbalance and displacement of population. Cash crops are being increased while the area under staple crops is being decreased. The USA, the IMF and the World Bank are providing loans in large quantities on the one hand, and the distrust and dissatisfaction among the population is ever increasing on the other. If the people began to distrust their government and their governmental machinery, who then can prevent country going disarray? Any ways, this is precisely what the IMF and the World Bank want!

Commenting on the World Bank's World Development Report 2000/2001 which is full of prescription for poverty reduction programmes, an Editorial in the Times of India stated that[20]:

> But it seems that an organisation, which has supported statism in the developing world since its inception—the World Bank is an architect of, failed policy—is loath to admit it.

Professor T.N. Srinivasan[21] of the University of Maryland in the USA, while commenting on World Bank's "*Strategy for poverty reduction*" simply wrote that "*...The bank has wandered into areas for which it has neither a mandate nor competence*".

B. CURVE NO. 2: REFORMIST APPROACH: OPPORTUNISM IN THE POST-INDEPENDENCE ERA SINCE 1947

Vasant Sathe, in his capacity as a senior minister in the Government of India in 1988, admitted[22]:

> "Anyone having open eyes can see that there is a serious erosion of credibility of the people in authority in the minds of the common man, in particular the youth. ... Serious trends that can lead the nation towards anarchy, combined with other divisive obscurantist and fissiparous force."

In order to get to power, the leaders struggling against the British colonialism in India, promised many social changes they would being about after Independence to remove the grinding poverty of the people. Jawaharlal Nehru, the first Prime Minister whose 17 years' rule was virtually unopposed and unchallenged, clearly implied so by publishing his *Discovery of India* in 1946, as it would be clear later in this chapter.

Once in power in 1947, these leaders did plan and carry out some of the promised changes, but only for a short while (distance OB). They did nothing of lasting value as far as the plight of the majority of people living below the poverty line is concerned, particularly in rural areas.

The leaders begin to indulge in pursuits of their selfish ends and personal glory and ego-gratification (distance BJ).

(At point J), after a shortwhile, when their authority begins to dwindle and threatened, the leaders again begin certain reforms (distance JK) until the discontent in the populace subsides and the opponents pacified through horse-trading in most cases, and by sheer brute police force in others — as the case may be.

Curve No. 2 in the diagram shows the reformist approach. This is the policy largely in practice in the post-Independence era, a policy to eradicate poverty being wholly opportunist.

Now we explain, with the aid of curve No.2, that the policy in the post-Independence era has been wholly opportunist.

On this curve No. 2, if we locate any four arbitrary points B, J, K, and L, we have three sets of interpretations from these four arbitrary points.

Before coming to power, the leaders struggling against the British colonialism, in the case of Nehru, it was a "tryst with destiny[23]" to uplift the level of the poverty-stricken masses. The leaders put forward their manifestos' outlining the social changes they would bring about when Independence would be achieved. Once in power (point O), they remain sincere only for a short while (distance OB) and immediately thereafter begin to circumvent the real issues facing the nation. They begin to indulge in the pursuits of their selfish ends and personal glory and ego-gratification (distance BJ).

From point B, let us now drop two coordinates BE and BH. We notice that the distance OE is somewhat greater than the distance OH. This means that in a reformist approach towards eradicating poverty, the amount of time taken (distance OE) corresponds to nearly similar achievements in social change (distance OH). We also notice that curve No.2 follows a cyclical pattern after distance OB. This means that in a reformist approach, the policy planners work only up to the point that satisfies their selfish ends, or when their authority and rule begins to dwindle and threatened. Hence the reformist approach is largely opportunist.

Once the selfish ends of the policy planners are satisfied (at point B), they begin to follow a decadent life and no longer work for the general welfare of the population until their own interests and positions are at stake again.

We see from Curve No. 2 that during certain times (from B to J), nothing is accomplished. Time is simply marked up. However, at point J, when the positions of the rulers begin to be threatened, they immediately begin certain reforms (curve rises from point J to K). From this distance (J to K on the curve), we notice that the time taken to accomplish social change is very little. When the discontent in the population subsides and the opponents pacified either by horse-trading or by brute police force (point K), the rulers again revert to doing nothing for social change, until the time when the situation takes violent turn again (at point L).

Thus the reformist rulers in India in the post-Independence era have been ruling by initiating certain social reforms whenever their positions have been threatened. No sincere attempts so far have been made to eradicate poverty except to keep the rulers being in positions of power and authority.

In nearly five decades after the Independence, all the reforms have meant more for strengthening political power rather than the common good of the wretched masses in general, where approximately 10 per cent of the population at the top level has been the main beneficiary of Independence in India, as concluded in Chapter 1 of this book.

So far, therefore, no genuine attempt has been made to eliminate the economic serfdom, illiteracy, and ignorance so widespread in rural areas, where over 80 per cent of the population dwells. What Mahatma Gandhi said in 1939 remains largely unaccomplished[24]:

> "In the case of the Indian villagers, an age-old culture is hidden under an entrustment of crudeness. Take away the encrustation, remove his chronic poverty and his illiteracy, and you have the finest specimen of what a cultured, cultivated, free citizen should be."

In the post-Independence era, a number of programmes had specifically been designed, developed, and implemented for rural development. During the VIth and VIIth Five-Year Plans (1980-90), massive investments have been made in Integrated Rural Development Programmes and (IRDP) and National Rural Employment Programmes (NREP). All of these programmes have not been able to scratch the surface of poverty in the rural areas. There are many reasons for their failures, but corruption, malpractices, and administrative inefficiency are most widely acknowledged. Vasant Sathe, a Cabinet minister in the Government of India for over a decade, gracefully acknowledged in 1988 that[25]:

> "It is a sad commentary on our image that the most popular films and stories are those which caricature and depict men in authority, including judges, as being utterly corrupt and without any character."

There is clearly a lack of good governance, which is largely attributed of the failures of the programmes of rural development for alleviating poverty. Dhiraj Nayyar[26] wrote that:

> Can an improvement in governance brought about by constitutional reform assist in the task of poverty alleviation, a task on which successive governments have failed to deliver for the last 53 years? The paper addresses

> this question and suggests that the answer is in the affirmative.
>
> ...There is the need to decentralise by devolving both functional and financial autonomy to panchayats in areas like education, health, roads and innovative programmes like watershed development. Second, there is a need for greater accountability and transparency in the functioning of government institutions. The lack of good governance according to me is an important reason behind persistent poverty. Economic development may be a more difficult task and there may be more to it than just governance but to ensure two proper meals for the entire population is not a Herculean task if efficient governance is provided.

While reviewing the book by Sundeep Waslekar, "*South Asian Drama: Travails of Misgovernance*", Professor K.L. Kamal[27] wrote that:

> ... "in all South Asian Countries, governance by personal rule has led to subjugation of the institutions of state to certain individuals and their coteries at different levels. As a result, the ability of the state to govern has grossly declined."

This is what precisely has been hallmark of achievements of more then half-a-century of Independence in this reformist approach aimed at poverty alleviation in all South Asian Countries.

C. REVOLUTIONARY APPROACH: THE PLIGHT OF THE POOR AND EXPLOITED

Bhanu Pratap Singh, former minister in the Government of India, dared to say[28]:

> "The only course open to us is to tear down the existing apparatus of politico-economic feudalism, which has been built up over the years in the name of socialism, but under

which only politicians and their capitalist friends have prospered."

Curve No. 3 in Fig. 6.1 shows the revolutionary approach. This is the policy generally adopted by the leaders who are filled up with passion to break the existing order and establish genuine efforts to uplift the lot of the poor and exploited.

If we locate on Curve No. 3 any arbitrary point C, and from it drop two coordinates CF and CI, we notice that distance OF is much lesser than OI. This means that, in a revolutionary approach towards eradicating poverty, it takes much less time (distance OF) to accomplish more social changes (distance OI).

However, the question now arises; long we would want to delay, how much time do we need to eradicate poverty? It depends upon the degree of poverty and ability and resolve and tenacity of purpose in the leadership cadres. In the case of the USSR, when Joseph Stalin came to power after the death of V.I. Lenin, he was too impatient to wait. Stalin set out towards the most brutal, ruthless and cruel methods to eradicate poverty from his nation. In his zeal to drag his nation from what was a most backward nation in Europe to the ultra-modern and most powerful, militarily and economically, nation on earth, challenging almost single-handedly, the imperialism of the USA and the West European nations and the United Kingdom.

Mahatma Gandhi said in 1937 that[29] "I would be the last to minimize the achievement of Russia, but the whole structure is based on force and violence". Professor Adam Ulam, of the Russian Research Centre at Harvard University of USA, widely acknowledged as one of the most authoritative biographers on Stalin[30], wrote:

> "A recognition of one undisputed achievement of Stalin's reign; from the weak third-rate industrial and military country which she have been in 1924 (when Stalin came to power after the death of Lenin), Russia had become one of the world's two superpowers."

Stalin himself said as early as in 1926, barely three years in power and far from the absolute dictatorship with which he ruled the erstwhile Soviet Union from 1924 till 1953 when he died[31]:

> "The main task of the bourgeois revolution consists in seizing power and making it conform to the already existing bourgeois economy, whereas the main task of the proletarian revolution consists, after seizing power, in building a new, socialist economy."

When the British were out and India became independent, Nehru's much-publicised ideals of building a socialist India based on Westminster type parliamentary democracy, as we noticed in earlier chapters, turned out to be a nightmare. The gap between the rich and poor was ever-increasing in the country as a whole, and the number of people below the poverty line in rural areas had always been increasing as we noted in Chapter[32] 1 (depending of course who is "defining" the poverty line and with what criteria).

In the case of China, the violence adopted by Mao was not comparable to that exercised by Stalin in the USSR; but so the achievements of Mao in China were not many as those of Stalin in the USSR. In Cuba, in her struggle towards socialism after the overthrow of the Batista tyranny, following the Marxist-Leninist approach with a tilt towards the USSR, the achievements in the fields of education, health, and social welfare have been remarkable. The overall economic structure is in much better shape than of any country of comparable size in the Caribbean region or any of the republics in Latin America. But we do not know of any evidence of force or violence in Cuba in her revolutionary approach to eradicate the poverty under the leadership of Fidel Castro.

One more interesting case that I would like to mention is that of Tanzania. After her independence from the British colonialism in 1962, five years later (in 1967) it was discovered by the leadership in that country that the main beneficiary of the fruit of independence had been the very same people who were already better-off even in the British colonial rule; and that the masses in general, in rural areas in particular, where almost 90 per cent of the population lived, were not the beneficiary of the fruit of Independence. Accordingly, the Arusha Declaration (1967) evolved a new policy to help the poor in rural areas, and a greater dependence on self-reliance than on foreign financial assistance. No violence or force has been used against the citizenry, and all the citizens are being expected to contribute the task of eradicating poverty on a 'voluntary basis'. So far we do not know of any concrete social (health and education) and economic achievements, but the approach in Tanzania under the enlightened leadership of President Julius K. Nyerere is indeed being appreciated, and closely watched, all across the world. President Nyerere has written extensively on the problems of development in countries, like his Tanzania, who have emerged after the disintegration of British colonialism. His writings are widely read and seriously debated.

In the case of India, we always have had a policy of nation building on the socialist pattern. But India is being governed on pure opportunism; about ten per cent of the people have been the sole beneficiary in the post-Independence era. About the socialism of Jawaharlal Nehru, let us quote no less a person than Jayaprakash Narayan[33]:

> "This is a most amazing Jawaharlal Nehru. ... Any normal individual is supposed to practice what he believes, but not the Prime Minister of India. He believes in socialism, but practices God knows what, certainly not socialism."

Jawaharlal Nehru ruled for seventeen years right from Independence in 1947 till his death in 1964. Indira Gandhi, the daughter of Jawaharlal Nehru, ruled India for fifteen years, from 1966 to 1977 and again from 1977 to 1984. Indira Gandhi ruled with the declared slogan of 'Garibi Hatao' (or Remove Poverty). But the way she ruled is well summed up in an editorial in a highly prestigious national newspaper of Canada, the Globe and Mail[34]:

> "The Chief Ministers in States controlled by Congress (I) have been handpicked Gandhi loyalists, their sycophancy matched, in many cases, by their venality and incompetence. ... She could revitalise the withered organization...(But)...for Nehru's daughter the priority is to perpetuate the dynasty."

After the death of Indira Gandhi in 1984 Mr. Rajiv Gandhi, the son of Indira Gandhi, became the Prime Minister and remained in power till 1989 when he did not win in General Elections.

In a revolutionary change towards the development of nation, particularly the economic development and the eradication of poverty and rural reconstruction, it takes much less time (distance OF) to accomplish the much-needed social changes on an unprecedented scale (distance OI).

Although we in India, ever since we rid ourselves of the British colonial yoke, have always had the declared policy of socialism to carry out the task of nation-building. How much "socialism" we actually practice, and not preach, is quite obvious from Chapter 1. (Economic Inequality).

India is one of the largest democracies in the world. When horse-trading goes on in "buying and selling" of the elected members of the legislatures and the Parliament, democracy becomes nothing but a joke, a mockery, a bourgeois democracy. The widely circulated daily newspaper in Canada, the Toronto Star[35] of May 30, 1982, commenting on elections held in several States of India, when Indira Gandhi's party failed to get the majority, commented: "Indira Gandhi buys loyalty of legislators (at about Rs. one million each)". In this context, it is quite appropriate to quote what V.I. Lenin (the Russian revolutionary leader, who just before the Revolution overthrowing the dreaded Czars) said in 1917, which is equally true of Indian democracy since the early 1970s to this day (1990s)[36]:

> "To decide once every few years which member of the ruling class is to repress and crush the people through parliament—this is the real essence of bourgeois

> parliamentarian, not only in parliamentary constitutional monarchies, but also in most democratic republics,...This democracy is always hemmed in by the narrow limits set by capitalist exploitation, and consequently always remains, in effect, a democracy for the minority, only for the propertied classes, only for the rich."

Thus the fruit of Independence is being enjoyed, as we have noticed in the preceding chapters, by those to whom the British handed over the reign of the nation. The philosophy of India, the essence of which is the Bhagwad Gita, calls for the *'niskama karma'*, or the desire less action. But the *karmas* of the rulers of India in the post-Independence era, particularly since 1970s, have been to satisfy their own selfish ends.

About the sorry plight of the poverty-stricken masses in rural areas and the lack of sacrifice on the part of those in power, R. Venkataraman, in his Presidential Address to the nation on the eve of Independence Day on 14-8-1990 admitted[37]:

> "...Gandhiji's aim was not the mere replacement of the Union Jack by the Indian Tricolor or the substitution of the white rulers by the brown...
>
> One-third of the population of India still lives at the subsistence level. The contrast between rich and poor, urbanite and villager, educated and illiterate, urban street-child and the girl-child in particular, suffers from painful disabilities.
>
> The vast mass of our people have remained outside the periphery of our progress. The distinguished economist Prof. Amartya Sen has recently said and I quote:
>
> "India's "self-sufficiency" in food has to be assessed in the light of the limited purchasing power of the Indian masses. There has been no great shortage in the market, no crisis to deal with, but at least a third of the rural population has regularly—and quietly—gone to bed hungry and malnourished."

...Unfortunately after the Independence the very concept of sacrifice for the sake of the country has disappeared from our lexicon."

Millions and millions of the people of India subsist below the poverty line, a shame in itself, on mere Rs. 100 to 120 per month, but the rulers of India comfortably live in mansions vacated by the British colonial rulers. As stated in Chapter 1, approximately 50 per cent of the population, *i.e.* over 400 million people still live below the poverty line, even nearly five decades after the Independence. Bhanu Pratap Singh, former Agricultural Minister of India, rightly warns that we have to[38]:

> "Tear down the existing apparatus of politico-economic feudalism, which has been built up over the years in the name of socialism, but under which only politicians and their capitalist friends have prospered."

Whether this would happen through peaceful means or violent upheaval, is to be seen; but, without its tearing down, the removal of poverty from the country is nothing but a mirage.

The so-called economic reforms and liberalisation of the economy that began in 1991, have not at all benefited the ordinary citizen among the teeming millions in India, and in any of the South Asian Country. The two polls carried out by the Times of India and the Economic Times reflect the opinion of a section of the society who are net users, who overwhelmingly falt that the reforms did not benefit ordinary citizen:

(a) *Times of India*[39]:

Poll Results:
Question: Has economic liberalisation failed to bridge the rich-poor gap in India?
Yes = 71%, No = 25%, Can't say = 4%
8215 Votes in all

The poll reflects the opinions of Net users who chose to participate, and not necessarily of the general public.

(b) *Economic Times*[40]:

Poll Results:

Question: Has the government lived up to the expectations of the common man in the last 10 years of reforms?
Yes = 21%, No = 75%, Can't say = 4%

This poll reflects the opinion of only those internet users who have chosen to participate.

As far as the interest of the poverty stricken mass are concerned, an eminent journalist of India Prabhas Joshi[41] wrote about the out come of the reforms:

> There has been many fold increase in the capital of the big industrial houses, and all of them have formed partnership with foreign companies. Some percentage of middle class and upper middle class got consumer goods much easily. The role of peasants, the artisans, and the small-scale entrepreneurs has been marginalised in these nine years of globalisation.

The net impact of these reforms on the economy in well summed up Mohan Guruswamy[42], an economist and former adviser to the Finance Minister of India:

> Mohan Guruswamy, the sad face of liberalisation: After almost 10 years, all key indicators tell us that the New Economic Policy of 1991 has decelerated India's growth, *Indian Express*, April 11, 2000, p. 11 (Former adviser to the Finance Minister).

- Industrial growth in 1981-91 was 7.7 per cent. Since 1991-92 it has been 5.8 per cent.
- Agricultural growth, which averaged 4.04 per cent annually for the decade 1981-91, fell to 2.3 per cent annually for the period 1991-92.

- The annual electricity generation capacity growth fell to 3.5 per cent during 1991-99 as opposed to 12.4 per cent for 1981-91. Electricity generation growth fell from 17 per cent to 7.8 per cent for the corresponding period.
- Trade deficits have been growing exponentially. In 1980-81 the adverse balance was Rs. 5,838 crore. In 1990-91 this was Rs. 10,645 crore. In 1998-99 it swelled to Rs. 55,478 crore.
- Poverty levels have remained constant at about 33 per cent. Therefore the numbers living below the poverty line have swelled by 46.7 million to 326.04 million in 1999.
- Rupee: in 1981 the Dollar was worth Rs. 8, in 1991 this changed to Rs. 18, and now to almost Rs. 44.

Many programmes of rural development aimed at alleviating poverty are being implemented by State Governments by themselves and with the assistance from the Central Government of India such as:

- Panchayati Raj
- Jawahar Gram Samridhi Yojana
- Swarnjayanti Gram Swarozgar Yojana
- Employment Assurance Scheme
- Rajiv Gandhi National Drinking Water Mission
- Indira Awas Vikas Yojana
- Credit-cum-Subsidy Scheme for Rural Housing
- National Social Assistance Programme
- Central Rural Sanitation Programme
- Drought Prone Areas Programme
- Desert Development Programme
- Integrated Wastelands Development Programme
- Land Reforms
- Council for Advancement of Peoples Action & Rural Technology
- DRDA Administration
- Training

A World Bank Country Study, *"India: Reducing Poverty and Accelerating Development"*, published in 2000 stated that India's anti-poverty programmes (APPs) *"are still a long way to go on ground"*[43]:

"Reforms in India's Anti-Poverty Programmes:

> India's anti-poverty programmes (APPs) are mainly run by the Central Government. They amount to some 6-7 per cent of total GOI budgetary expenditure, or 1 per cent of GDP. Even accounting for inflation, they have been growing at 10 per cent per annum since 1992-93. There are three main types of APPs : rural works, self-employment, and food subsidy programmes. All three have been subject to reform in recent years.
>
> Food subsidy programmes make up about 55 per cent of total APP spending. By far the largest food subsidy programme is the PDS, which was explicitly targeted towards the poor at the national level in 1997 and renamed the Targeted PDS (TPDS).
>
> Self-employment programmes make up only about 5 per cent of total APP spending, but have received a lot of publicity, most of it bad, on account of the poor performance of the Integrated Rural Development Programme (IRDP). This year the IRDP was scrapped, along with five other small self-employment programmes, all of which were replaced by a single programme—the Swarnjayanti Gram Swarozgar Yojana (SGSY).
>
> Rural works programmes account for about one-third of APP spending. There are two main schemes, the EAS (Employment Assurance Scheme) and the JRY (Jawahar Rozgar Yojana), now renamed the JGSY (Jawahar Gram Samridhi Yojana).
>
> The rationalisation and better targeting underlying the above reforms are both big steps forward. All the programmes are also giving a greater role to rural local

government for implementation and for beneficiary selection and monitoring. The reforms also lay stress on transparency, making information about the programmes public at the village level, and on the importance of physical, financial, and social audits. While these reforms are very welcome, there is still a long way to go on the ground.

For the various programmes of rural development for alleviating poverty, the Government of India allotted:

Rs. 7,517 crore in 1999-2000,
Rs. 6,760 crore in 2000-2001,
Rs. 8,197 crore in 2001-2002,
Rs. 7,973 crore in 2002-2003.

As reviewed and analysed in Chapter 1, despite plethora of these multi-dimensional programmes of rural development on the one hand, and large sums being spent to implement them on the other, poverty still remains as perennial a problem as ever before. So what is required is a revolutionary approach, a paradigm shift from:

(a) The evolutionary approach of the World Bank and the IMF in the post-Independence era who dictate the policies of the developing countries for removing poverty (or of the British Colonial approach which was in any case, not designed to do any good for the country, except to plunder it effectively), and
(b) The reformist approach being practiced in the last half-a-century since Independence, which has ended up in what K.L. Kamal[44] wrote that:

> ..."in all South Asian Countries, governance by personal rule has led to subjugation of the institutions of state to certain individuals and their coteries at different levels. As a result, the ability of the state to govern has grossly declined."

The political question in this revolutionary approach in

alleviating poverty is simple and brutal. To quote K.L. Kamal[45] again:

> "Who will lead the masses, how shall they get organised, how shall they fight the custodians of power who have all the resources at their command, what would be the nature of the institutions and structures that could be envisaged, what institutional restraints would be imposed on the future leaders of the society?

NOTES

1. K.N. Raj, "Poverty and Political Development," in *Economic and Political Weekly* (Weekly Magazine), Bombay; Annual Number, February 1977; p. 199.
2. K.L. Kamal, Book Review (Sundeep Waslekar: *South Asian Drama: Travails of Misgovernance*, Konark Publisher, New Delhi, 1997 in Administrative Change, Jaipur, Vol. No. XXV, July-December 1997, p. 80.
3. For a quantitative extent of poverty in India and the developing countries, see Chapter 1 in this book.
4. For details about productive work and gainful employment, see Chapter 2 of this book, "The Formation of an Agro-Industrial Complex".
5. Integrated Approach to Rural Development (New York: United Nations, 1977); pp. 56, 73 (paragraphs 319).
6. For detailed analysis, *see* Sures Chandra Jain, *Education for Socio-Economic Development*, Concept Publishing Co., New Delhi, 2004.
7. Panchayati Raj, a Hindi word, meaning "Self-Rule" by the villagers at the local levels on matters dealing with land disputes, or general everyday living problems that otherwise require courts, police, etc.
8. Leith Griffin, "Rural Development: Policy Options" in Edgar O. Edwards, ed., *Employment in Developing Countries* (New York: Columbia University Press, 1975); p. 182.
9. *Ibid.*, p. 184.
10. See Section A of Chapter 2—The Policy Issue of this book.
11. Integrated Approach to Rural Development, *op. cit.*, pp. 41-42 (paragraph 162).
12. For detailed analysis, *see* Sures Chandra Jain, *Education for Socio-Economic Development*, Concept Publishing Co., New Delhi, 2004.
13. Karl Marx and F. Engels, *On Colonialism* (New York, International Publisher, 1972); p. 83.

14. *Ibid.*, p. 82
15. The World Bank, World Development Report 2000/2001, Oxford University Press, New York, 2000, pp. 6-7.
16. *Ibid.*, p. 195.
17. *Ibid.*, p. 179.
18. *Ibid.*, pp. 21-23.
19. अखिलेख अखिल, तीसरी दुनिया के देशों पर साम्राज्यवादी आक्रमण, *दैनिकभास्कर*, भोपाल, दि. 5-3-2000, पृष्ठ 1 (रसरंग).

    The above passage is translated from the following original in Hindi:

    आज भारत पर सबसे ज्यादा नजरें बैंक कोष की ओर लगी हुई हैं । विश्व बैंक कोष की मदद से यहाँ भी नदी घाटी परियोजनाएँ चल रही हैं, लोगों का विस्थापन जारी है पर्यावरण असंतुलन बढ़ रहा है, नकदी फसलों को बढ़ाया जा रहा है, जबकि खाद्यान्न फसलों का क्षेत्रफल घटता जा रहा है । एक तरह से विश्व बैंक, कोष व अमरीका जैसे देश बड़ी मात्रा के ऋण मुहैया करा रहे हैं, जबकि दूसरी ओर भारत की आम जनता में असंतोष, अविश्वास बढ़ता जा रहा है और जब देश की जनता का सरकार और व्यवस्था के प्रति अविश्वास बढ़ता गया तो फिर देश को बर्बादी से कौन बचा सकता है । आखिर बैंक कोष चाहता भी तो यही है ।
20. Editorial, Go Home World Bank, *The Economic Times*, New Delhi, September 29, 2000, p. 12.
21. T.N. Srinivasan, WDR 2000-01: A poor report on poverty, *The Economic Times*, New Delhi, September 28, 2000, p. 7.
22. Vasant Sathe, "Refusal to Think: An Act of Cowardice", in *Yojana*, New Delhi, January 26, 1988; p. 12.
23. This phrase, "Tryst with destiny", was used by Nehru on August 15, 1947, when the British Governor-General, Lord Mountbatten, handed over power to Nehru.
24. Mahatma Gandhi, *Collected Works*, Vol. 68 (New Delhi: Ministry of Information and Broadcasting, 1977); p. 266.
25. Vasant Sathe, "Refusal to Think: An Act of Cowardice", in *Yojana*, New Delhi, January 26, 1988; p. 12.
26. Dhiraj Nayyar, Alleviating Povery: Role of Good Governance and Constitutional Reform, *Economic and Political Weekly*, Mumbai, October 14, 2000, pp. 3739-41.
27. Sundeep Waslekar, "South Asian Drama: Travails of Misgovernance", Konark Publishers, Delhi, 1997. Book Review by Professor K.L. Kamal in "*Administrative Change*", Jaipur, Vol. XXV, No. 1 July-December, 1997, p. 80.
28. *Times of India*, New Delhi, 20-3-1989; p. 3.
29. Mahatma Gandhi, *Collected Works*, Vol. (New Delhi: Ministry of I&B, 1976); p. 266.
30. Adam Ulan, *Stalin: The Man and His Era* (New York: Viking Press, 1973); p. 8.

31. Joseph Stalin, *On the Opposition* (Peking: The Foreign Languages Press, 1976); p. 276.
32. So is the progressive increase in the absolute number of illiterates despite a decrease in the percentage of literacy. See the author's other book, Education for Socio-Economic Development.
33. Jayaprakash Narayan, *Towards Total Revolution*, Vol. II (New Delhi: Popular Prakashan, 1978); p. 186.
34. The Globe and Mail, Toronto, Canada, 8.2.1983; p. 8.
35. Toronto Star is a widely circulated daily newspaper of Toronto, the largest city in Canada.
36. V.I. Lenin, *State and Revolution* (Moscow: Progress Publishers, 1967); p. 79.
37. Quoted from the text of the speech of the President of India, *The Hindustan Times*, New Delhi, 15-8-1990; p. 4.
38. *Times of India*, New Delhi, 20.3.89; p. 3.
39. *Times of India*, New Delhi, 16-11-2000, p. 1.
40. *Economic Times*, New Delhi, 24-2-2000, p. 1.
41. प्रभाष जोशी, *जनसत्ता*, 20-5-2000, पृ. 6.

    The above passage is translated from the original in Hindi :

    भूमण्डलीकरण एवं उदारीकरण के 1991 से पिछले 10 वर्षों में 2000 तक निम्नलिखित परिणाम सामने हैं:

    1. बड़े औद्योगिक घरानों की सम्पत्तियों में बेतहाशा वृद्धि हुई और इन सबने विदेशी कंपनियों से गठजोड़ कर लिये,
    2. मध्यम वर्ग के कुछ अंश एवं उच्च मध्यम वर्ग को उपभोक्ता सामग्री आसानी से प्राप्त हुई, एवं
    3. ''9 बरस से भूमण्डलीकरण में देश के किसानों, कारीगरों और छोटे-छोटे उद्योग चलाने वालों को किनारे लगा दिया है । औरतों और दलितों की हालत और बिगड़ी है ।''
42. Mohan Guruswamy, The sad face of liberalisation: After almost 10 years, all key indicators tell us that the New Economic Policy of 1991 has decelerated India's growth, *Indian Express*, April 11, 2000, p. 11 (Former advisor to the Finance Minister.
43. India: Reducing Poverty and Accelerating Development (A World Bank Country Study) Oxford University Press, New Delhi, 2000, p. 19.
44. Sundeep Waslekar, "South Asian Drama: Travails of Misgovernance", Konark Publishers, Delhi, 1997. Book Review by Professor K.L. Kamal in *Administrative Change*, Jaipur, Vol. XXV, No. 1, July-December, 1997, p. 80.
45. *Ibid.*, p. 81.

Index

1

The Problem: People Below the Poverty Line

"A violent and bloody revolution is a certainty one day, unless there is a voluntary abdication of riches and the power that riches give and sharing them for common good."

—*Mahatma Gandhi, 1941*

"Despite high economic growth in some parts of Asia, increasing inequality both within and across countries has meant that poverty has remained a persistent problem."

—[illegible]

Poverty is essentially a relative phenomenon. Poverty has different meanings for different segments. I would therefore first of all like to define what I mean by poverty. To my mind, as far as the present socio-economic milieu of India is concerned, a person is in a state of poverty when he or she is not in a financial position to afford the following five basic requirements of a dignified human life:

- Sufficient food
- Minimum hygienic housing
- Adequate clothing
- Access to primary education
- Essential health care.

With anyone of the five requirements lacking, a person is living in poverty. To afford the basic amenities of life, a person